Creative
Sciencing
a practical approach

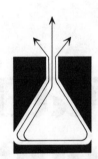

ALFRED DeVITO
purdue university

GERALD H. KROCKOVER
purdue university

Creative
Sciencing
a practical approach
second edition

little, brown and company **boston   toronto**

PREFACE

As science educators, we have asked both preservice and in-service teachers why science is not always taught in a regular, sequential manner throughout grades K–8. Often the responses have been "We don't have the time," "We don't have the materials," and "We don't have an adequate background in science." This book is intended to help you develop the knowledge, skills, and teaching strategies needed to become a creative, innovative teacher who can present science lessons in an exciting way to a variety of children in a variety of situations.

Creative Sciencing: A Practical Approach, Second Edition, is designed to provide preservice and in-service elementary school teachers with a practical guide for science instruction. This book may be used as the basic text for the elementary science methods course, or it can be used with an activities book such as *Creative Sciencing: Ideas and Activities for Teachers and Children* or with an array of other materials.

Section 1 explores the role of science in the curriculum, major science programs, and the selection of an elementary science program.

Section 2 examines the importance of changing your attitudes and practices, developing questioning techniques, using the laboratory in science, and individualizing instruction.

Section 3 presents methods and techniques for teaching children to teach themselves, including task cards, puzzler activities, audiotutorial tapes, invitations to investigate, skillettes, science-book kits, the buzz box, science games, modules, and interest centers.

Section 4 treats science as part of the integrated curriculum. It includes ideas and activities for science and mathematics, metrics, art, social science, social studies, health, language arts, and reading.

Section 5, "What They Neglected to Tell Me," features practical approaches for meeting such challenges as maintaining order in the classroom, working with gifted children, mainstreaming, teaching controversial issues, using museums, and selecting textbooks.

Section 6, "Creative Sciencing Evaluations," examines evaluation in all three domains of learning — cognitive, affective, and psychomotor. A selection of evaluation ideas and techniques are included for use by pre-service and in-service teachers.

A sample science process assessment is found in Appendix A, and Appendix B lists sources for more ideas and activities. An instructor's manual with objectives and teaching hints is available.

This second edition of *Creative Sciencing: A Practical Approach* has been expanded by nearly forty percent. New material is found in Sections 1, 2, and 6. Section 5 is a major new section. The nature of these additions and changes were strongly influenced by the advice we received from our colleagues. A questionnaire was sent to a sample of sixty colleges and universities that used the first edition. We sincerely appreciate the suggestions made by users of the first edition. We would also like to thank Dr. Shirley Brehm, Michigan State University; Dr. Lowell Bethel, University of Texas at Austin; and Dr. Delmar Janke, Texas A & M, for their excellent work in critiquing the revised manuscript. We are indebted to Barbara De Vito for her typing of the manuscript and to Sunny Lou Shulkin for editing our manuscript.

Alfred De Vito

Gerald H. Krockover

CONTENTS

science in the integrated curriculum
Great risks, greater profits   131

what they neglected to tell me about...
Does it matter?   151

creative-sciencing evaluations
Evaluation raises hell with trust!   197

Appendix A:

Appendix B: Where do we go from here?

1
Setting the stage

You can make the difference

It is in fact nothing short of a miracle that the modern methods of instruction have not yet entirely strangled the holy curiosity of inquiry; for this delicate little plant, aside from stimulation, stands mainly in need of freedom; without this it goes to wrack and ruin without fail.

— Albert Einstein

line up!

"OK children, close your reading books. It's time for recess! If you line up quietly and behave during recess, we *may* have science class this afternoon from 1:00 to 1:30.

"Row one may line up. Row two, if you don't quiet down, you may not be able to get out today. Row three may line up. Row four. Row two, I'm glad that you're now ready."

next, recess time —
next, mathematics workbook time —
next, lunch time —
next —

"When William gives me his attention and closes the book he is reading (*Common Rocks of North America*), we can begin science. You children are very lucky. When I went to school, I never had science until junior high or high school. My elementary school teachers said that they didn't have time to teach science, and besides, we had to spend all our time learning to read and do arithmetic. Occasionally, however, other students in the same school had science in the afternoon for one-half hour, two days a week. Aren't you lucky that you can have science three days a week? Where were we? Oh yes, William, close that book. Now let's all open up our science books to page 31. We will have a movie tomorrow showing the bones of the skeleton. I want you to look at that diagram and memorize the names of those bones so that you will be ready for the movie tomorrow. Also, we will have a quiz Friday, and if you miss more than two bone names, you will get an F. This will mean a makeup quiz on Monday instead of recess. Aren't you lucky that you can have science three days a week?"

What we say and *what we do* can have a lasting effect upon children. Whether we wish to admit it or not, we convey our likes and dislikes, our strengths and weaknesses, our values and opinions, and our prejudices to our students.

Two students who are now in college have these comments to make about science teaching. Could either of them be our former student?

Betty: When I think of past approaches to science, I can only be reminded of the horrible experience I had when I was in elementary school. We had a science lesson once a week. It would usually be on a Friday afternoon. For half an hour, we would take turns reading from a science book. We would never do any of the experiments we read about. Then, at the end of each section, several questions would arise concerning the material covered. If the class was confused or did not understand a point, the teacher would never discuss the problem. Instead, that question was assigned as homework due the next day. The teacher would hand back our papers with next to no discussion on the issue. All through elementary school, I always had the feeling that my teachers' favorite subject was anything but science. In one of my classes, we used to have a choice of either art or science. Needless to say, art would always win.

Sandy: I feel that today's children are basically no different from myself as I was twelve years ago. I was always bored with science class, thinking there was nothing worse than memorizing facts about something you could not even visualize. I actually thought that was all there was to science because, from the time I entered first grade until high school, I was only acquainted with the memorization of facts and maybe one or two demonstrations per year carried out by the teacher. The other students and I were not presented with any alternatives; therefore, we knew of nothing better and swallowed what the teachers offered.

Setting the Stage / 3

SCIENCE AND THE CURRICULUM

Science is a human undertaking. It is the process of seeking explanations and understanding of the natural world. It also includes the knowledge that the process produces. To the extent that it is a human undertaking, its inclusion in the curriculum is mandated. To the extent that the attitudes and ideals of science are valuable and lasting throughout students' lives, education should reflect the spirit of science. It would be a courageous venture to have science serve as the spine of the curriculum.

Creative sciencing is more than just good science: it is good reading, language arts, social studies, mathematics, art, music, and physical education. Most importantly, it is good education. Creative sciencing is the approach that advocates the integration of all areas of the curriculum with the attitudes, ideals, and spirit of science.

An awareness of the attributes of science dictates that educators assign to science a larger and more explicit place among the goals of education. This approach should evidence itself in a classroom atmosphere that reflects scientific ideals as they permeate other areas of the curriculum. This is not a new concept. Good teachers have been doing this for generations. Integrating the philosophy of science into the total curriculum is simply good teaching.

Critics of this concept claim that integrating the various areas of the curriculum will eventually reduce literacy and lessen children's mastery of skills and facts. Such a concern cannot be passed off lightly. But is this claim legitimate? Dissecting a frog to study the component parts in isolation, devoid of life, is not the same as studying them as functioning parts of the total organism. Each approach undoubtedly has advantages. When the isolated approach is desirable, topics such as mathematics can be taught as clean, uncluttered mathematics. And, when the integrated approach is necessary, mathematics should be taught as it weaves and threads itself through science, art, music, social studies, or language arts.

Science is rooted in observation. Science is a search for knowledge through experimentation; a search for knowing and understanding; a questioning of all aspects of the environment; and the collection and analysis of data and the interpretation of their significance. Science has a respect for logic and leads to progressive confidence in one's ability to draw conclusions based on one's own authority. Science develops a spirit of inquiry. Some scientists assume that all things are connected, all things serve a function, and that there is an explanation for everything.

The acquisition by students of the attributes and values of science would be highly desirable and well worth the effort.

A good way to permit the spirit of science to be reflected in all of education is to join the processes of science to the content of specific areas of the curriculum. These processes are observing, classifying, measuring, inferring, predicting, formulating hypotheses, interpreting data, and making models to explain the evidence. These processes coupled with an inquiring approach to learning should produce creative, innovative, independent, thinking students.

Many topics comprise what is called the curriculum, among them: reading, spelling, social studies, language arts, science, mathematics, music, physical education, and art. Who decided these are the topics that should be taught? Who decided spelling should be taught as a separate topic? Or science? Who decided elementary school teachers should teach *all* subjects and secondary school teachers *one* subject? Who decided that departmentalization should not begin until the fourth grade? And, most important, who decided that science should only be taught in the early grades once a week for thirty minutes; in the intermediate grades, two or three times a week for forty-minute periods; and, in the secondary grades, five days a week for fifty-minute periods?

In many elementary schools the presentation of science in the primary grades is nonexistent. Surely, there are exceptions to the order of the elementary teacher's topic-teaching preference. Numerous teachers do teach science at the lower elementary levels but, unfortunately, not enough do. One study indicated that only 16 percent of a sample of 102 primary classroom teachers taught science.[1] The percentage of schools that include science as a separate subject increases by grades. Most commonly, schools provide instruction in science two, three, or five periods per week. The percentage of schools teaching science five periods per week increases in the upper grades. This may be the result of departmentalization in the intermediate grades.

Science is an undertaking involving speculation. It is also continuous. It is from this quality of speculation that the conceptual or "big" ideas in science originate. Such speculation is highly imaginative and depends on intuition and inspiration. This is the creative dimension of sciencing —

1. H. F. Fulton, R. W. Gates, and G. H. Krockover, "An Analysis of the Teaching of Science in the Elementary School: At This Point in Time — 1978–79," *School Science and Mathematics,* forthcoming.

the big idea rarely comes from an examination of the facts and the careful use of logic. The transition from a speculative idea to a fruitful theory depends completely on experience and observation. This experience and observation should start early in the primary grades.

Proponents of early and continuous instruction in science have stated numerous reasons for teaching science. Basically, there are three purposes: to familiarize students with a basic body of knowledge, to help students develop proper attitudes toward science and the world of technology, and to assist students in acquiring the fundamental skills of science so they will be able to function better within a scientific environment. All these are highly desirable goals; none should be neglected. It is sometimes difficult, however, to sell students on the fact that knowledge of science is valuable. A suggested approach is to stress that learning science is a highly personal thing.

Education has among its major objectives the development of each student into a creative, innovative, independent thinker. Creative sciencing can contribute to the improvement of students' self-concepts.

Science is creative. It does develop observational skills, skills in making relevant distinctions, skills in critical thinking, and skills in experimenting. The most rewarding and personal aspect of studying science is the joy of the intellectual power one experiences and develops as one engages in sciencing. The realization of the independent power one can bring to bear in the solution of problems is monumental. There is joy and excitement in having confidence in oneself, posing questions and problems, seeking answers and solutions, and making conclusions based on confidence in one's own investigation and quality of work. This experience transforms statements such as, "I believe because I read it" or "I believe because someone told me," to "I believe because the data I have garnered and interpreted lead me to believe so." The power that comes from the discovery of one's ability to carry on an investigation with confidence is one of the salient values of studying science and is not limited by grade level.

Could the letter on the next page have been sent out by your school district? Why? Why not?

EDSEL MEMORIAL HIGH SCHOOL
Anywhere, U.S.A.

August 1, 1980

Dear Parents of Our Graduates:

As you are aware, one of your offspring was graduated from our high school this June. Since that time it has been brought to our attention that certain insufficiencies are present in our graduates, so we are recalling all students for further education.

We have learned that in the process of the instruction we provided we forgot to install one or more of the following:

— at least one salable skill;
— a comprehensive and utilitarian set of values;
— a readiness for and understanding of the responsibilities of citizenship.

A recent consumer study consisting of follow-up of our graduates has revealed that many of them have been released with defective parts. Racism and materialism are serious flaws and we have discovered they are a part of the makeup of almost all our products. These defects have been determined to be of such magnitude that the model produced in June is considered highly dangerous and should be removed from circulation as a hazard to the nation.

Some of the equipment which was in the past classified as optional has been reclassified as

William C. Miller, "Recalled for Revision," *Phi Delta Kappan,* vol. 53 (December 1971), back cover. © 1971, Phi Delta Kappa, Inc. By permission.

Science and the Curriculum / 7

standard and should be a part of every product of our school. Therefore we plan to equip each graduate with:

- —a desire to continue to learn;
- —a dedication to solving problems of local, national, and international concern;
- —several productive ways to use leisure time;
- —a commitment to the democratic way of life;
- —extensive contact with the world outside the school;
- —experience in making decisions.

In addition, we found we had inadvertently removed from your child his interest, enthusiasm, motivation, trust, and joy. We are sorry to report that these items have been mislaid and have not been turned in at the school Lost and Found Department. If you will inform us as to the value you place on these qualities, we will reimburse you promptly by check or cash.

As you can see, it is to your interest, and vitally necessary for your safety and the welfare of all, that graduates be returned so that these errors and oversights can be corrected. We admit that it would have been more effective and less costly in time and money to have produced the product correctly in the first place, but we hope you will forgive our error and continue to respect and support your public schools.

Sincerely,

P. Dantic, Principal

THE BIG THREE AND THE TERRIBLE THREE

The basic purpose of education is to assist students to develop the ability to think. Numerous studies have supported this notion. Benjamin S. Bloom points out that it is possible for 95 percent (Yes, 95 percent) of our students to learn all that the school has to teach, all at near the same mastery level.[2] However, rather than meeting this goal, instruction in school tends to widen the gap between high- and low-achieving students. Unsuccessful students become more unsuccessful and successful students become more successful as they advance through each grade level.

Instilling in children the basic values of learning should be the goal of every school program, and we need to decide what learning categories are basic to education. Which of these categories do you think students would retain after one year? Rank the six categories listed from 1 (retained the longest) to 6 (the shortest).

My Ranking	Learning Category
———	*Factual Material*

Examples: Elements are made up of only one kind of atom. The speed of light is 300,000 kilometers per second.

| ——— | *Attitudes About Self in Relation to Subjects, Studies, and Others* |

Examples: I like science. I enjoy doing science activities. My teacher is a big bore.

| ——— | *Motor Skills* |

Examples: printing, typing, measuring, riding a bike

| ——— | *Conceptual Schemes* |

Examples: Moving air creates weather. Man is a product of heredity and the environment.

2. *Human Characteristics and School Learning* (New York: McGraw-Hill, 1976), p. 71.

_____ *Nonsense Syllables*

Examples: adc, bld, adv, ghk

_____ *Thinking Skills and Processes*

Examples: How to solve problems; how to observe, predict, infer, form hypotheses, communicate, classify — creatively

Check your answers! Research studies conducted over the past years have confirmed the order to be:

The big three

1. Attitudes about self in relation to subjects, studies, and others. (100% retention after one year)
2. Thinking skills and processes. (80% retention after one year)
3. Motor skills (70% retention after one year)

The terrible three

4. Conceptual schemes (50% retention after one year)
5. Factual material (35% retention after one year)
6. Nonsense syllables (10% retention after one year)[3]

Does the order surprise you? While the results of research indicate that science instruction should concentrate on the categories in the Big Three, concepts and facts should not be eliminated from the science-teaching repertoire. Science facts and concepts in isolation are of little value; they need to be used to develop the three higher learning categories.

points to ponder

— In what categories does most instruction in our schools take place: the Big Three or the Terrible Three? Why?
— In which category should most instruction in our schools take place? Why doesn't it?
— What are the implications of these results for the way that we plan learning experiences? Should experiences be planned for the Big

3. Data reported by L. J. Cronbach in *Educational Psychology* (New York: Harcourt Brace Jovanovich, 1963), p. 10.

Three or the Terrible Three? Cite reasons for your answers.
— If you teach for the Big Three, should you neglect the Terrible
Three? Explain your answer.

BE CREATIVE!

Science is a natural for developing the Big Three learning categories
(attitudes, thinking skills and processes, and motor skills). Science is
also a natural for creative teaching. Society's survival demands exploration, discovery, invention, and experimentation. These are activities of
the mind that involve curiosity, imagination, and a sense of destination.
Among the results of these mental activities are scientific theories, inventions, literature, and works of art. While these mental activities are commonly associated with science, creative students who have been encouraged to develop them will extend them to all facets of their lives.

Creativity, although deemed a valuable trait, is probably our least
developed talent; it is a thinking process; it is the power to recombine
things that already exist into something new. Creativity can also be
described as a succession of acts, each dependent on the one before, and
suggesting the one after. Creative recombinations can produce a new
object, a new technique for performing a familiar task, or a new solution
to a problem.

Creative people are problem seekers and problem solvers. As teachers
we must acquire and then promote creative talents in ourselves and our
students. Creativity comes naturally to some people; for most individuals, however, creativity is a do-it-yourself achievement. It is personal.
It appears most readily available when one is oneself. So, relax. Be yourself. Discover yourself. Many people have never taken time to become
acquainted with themselves. People who have discovered and accepted
the parameters of their personalities appear more free to concentrate on
developing their creativities. Here are some suggestions for involving
yourself in *your* creativity:

ask yourself questions

Don't always accept what you see or hear as final. Delve into each situation. Challenge yourself to come up with several searching questions
about any situation. Think of extending the scope of the problem or situation by asking provocative questions of yourself, of others, about the

situation. This should encompass a thorough understanding of the problem, some plan for actively embracing the problem, and a verification of results.

vary your perspective

When viewing a problem, look for ways of modifying it, substituting variables, rearranging components of the problem to arrive at new relationships, or occasionally reversing processes. Remember, a cow doesn't look the same viewed from opposite ends. Try magnifying components of the problem; minimizing components. What does this do for trying out recombinations or new relationships?

get involved

Creative spontaneity springs from fresh insights that mingle with old ideas and exit from the mind as new expressions. Keep fresh discoveries coming by getting involved in something — preferably something new. Become an expert in some area but keep reaching out for new experiences. Quiz people. Wonder about everything. Observe. Associate with people who exhibit the creative spark. Creativity rubs off. It is infectious.

live happily

Happiness with a pinch of frustration, stirred well with a germ of an idea, drenched in perseverance, is highly conducive to creative production.

take a chance

Become a calculated risk taker. Creativity demands daring personalities. Gamble with dreams. Creative people dream high and then engineer these dreams down to reality. Don't be bashful. Kick at stars. Run with the wind. Try something. Remember: if you think you are, you are. And, after all, you really are.

WHY TEACH SCIENCE?

Science is claimed by some people to be a "many splendor'd thing." It is often described by these people as fascinating, challenging, beautiful, interesting, and intellectually satisfying. But not everyone agrees: for many people, unfortunately, science is an unpleasant sortie into a vast sea of facts, formulas, concepts, vocabularies, laws, theories, and generalizations coupled with unfamiliar intellectual applications. So it is not unusual for students and, occasionally, teachers to ask the question, Who needs science (and why)?

The need to know the content of science does not answer the question or satisfy these questioners. If science is to gain stature in the eyes of those discontented with the science curriculum, it must be promoted not on the basis of the value of content acquisition but, rather, on the basis of what science can do to the individual's ability to think. There is no substitute for knowledge, the content of science. However, facts in and of themselves, while a vital and necessary component of the process of sciencing, are not a sufficient part of the process.

What is unique to the study of science?

Why should science be given equal time with other areas of the curriculum? While each of the sciences has its own particular approaches, tools, and methods for discovering and ordering information, all sciences have the process of experimentation in common. Experimentation is unique to science. Experimentation is the process of establishing the existence of something or finding the solution to a problem. Concomitant with the acquisition of the skill of experimentation, elementary school children should acquire from a study of science:

— the habit of questioning all things;
— ability to evaluate premises, impinging variables, and the consequence of anticipated action;
— the spirit of demonstrative replication and verification;
— a desire to search for data and their meanings;
— a respect for logic in the development of a strategy for inquiry.

If there is indeed a joy to sciencing, it is in the awareness and development of the intellectual power that each child is capable of marshalling towards the investigation and solution of problems. With the acceptance of this awareness and development as a primary outcome of sciencing, the learner will see the facts, concepts, formulas, vocabularies, laws, theories, and generalizations of science as justifiable and even desirable.

What should determine the elementary science curriculum?

Determination of the elementary science curriculum content is contingent on several factors — the goals of society, the structure of the sciences, and the goals of the learner. The goals of society are usually broad and pervade the entire curriculum, K–12. The science program, while seemingly well delineated, is often confounded by the nature and geographic mobility of the learners. A more detailed discussion of these factors follows.

The goals of society The structure of a democratic society demands that its citizens be responsible, competent, and ethical. Education in the sciences can assist children to understand the dependence of society upon scientific and technological achievement and to accept the fact that science is basic to living. Elementary and middle school children should be led to understand the relation of basic research to applied research as well as the interplay of technological innovations and human affairs.

The structure of the sciences Desirable outcomes of an elementary science curriculum should include:

— the acquisition of scientific knowledge;
— the communication of scientific knowledge;
— the use of scientific knowledge by children in the solution of problems;
— the means for survival;
— the personal fulfillment reached from applying science in the home, as a consumer, and, perhaps, as a vocation.

Education in the sciences should be structured to help individuals acquire skills and knowledge and to nurture their emotional adjustment to relate successfully to themselves and to the world around them. Elementary school children should be afforded a maximum science education so that they can best manage their personal and collective lives to survive in the face of change.

The goals of the learner Science education is the human enterprise of striving to explain natural phenomena. It should extend children's ability to learn, to relate, choose, communicate, challenge, and respond to challenges. Such abilities will enable them, as individuals, to live with purpose in the world of today and tomorrow, achieving pleasure and satisfaction in the process.

An elementary school science curriculum should present appropriate science content. Further, the curriculum should help students learn processes that will make them capable of observing with discrimination, of classifying observations of data, of quantifying their observations and communicating through the use of graphs, of synthesizing and modifying explanations, and of making and testing predictions against theory.

Content selection in the science curriculum

Over the past twenty years much that has happened in public school science instruction has been good. Numerous federally funded projects at both the elementary and secondary school level have pointed the way for what might be generally accepted as appropriate science instruction for public school students. Fallout from these projects has carried their spirit and intent into commercially prepared science texts, visual aids, and a wide assortment of hands-on equipment. Currently, a variety of good materials is available for science instruction.

Generally speaking, elementary school science should be exploratory. The nature of the learner rather than the structure of the sciences dictates the curriculum. The curriculum in the primary grades should concentrate on the development of the basic process skills of science — observing, classifying, measuring, communicating, inferring, predicting, and recognizing space/time relationships.

These skills should be developed around hands-on science activities that:

— stimulate curiosity and enthusiasm;
— capitalize on the creative nature of young children;
— foster continuous creativity;
— nurture the habits of systematic observations;
— initiate and sustain qualitative and quantitative thinking and representations;
— promote the development of scientific thought.

A science vocabulary, while important, should not be acquired at the expense of the acquisition of ideas. Ideas should precede the vocabulary of science.

The science curriculum content of the intermediate and middle school should focus on the enlargement and continuity of science skills. In these grades the curriculum should concentrate on content acquisition as a vehicle for the development of such integrated process skills as experimenting, controlling for variables, collecting and interpreting data, and building scientific models.

The vitality of science should be promoted by constantly presenting it as a challenge. Inquiry, the scientific spirit of searching for answers, should not be sacrificed to the tedium of science. This is not to deny that the tedium exists, but it is tolerated best when science's vitality is fully appreciated by the learner.

Curricula, like everything else, must change, but that change must be carefully monitored. Priorities must be delineated. Given an exponentially increasing body of knowledge, the science curriculum content must be organized to assign appropriate material to different grade levels and student groups. Physicists, science educators, and teachers, for example, need to agree on appropriate topics in physics for students in the elementary school.

The capstone to any science instruction should be the capacity of an individual to relate sensitively, to think divergently, and to perform imaginatively in confrontations with people and ideas.

For every complex problem there is usually a short, simple answer. Unfortunately, it is usually the wrong answer. The right answer is the one we want badly enough to work hard to obtain.

WHICH SCIENCE PROGRAM IS BEST?

decisions, decisions, decisions!

An addendum to the question, Which science program is best?, should be "for the learner, for you, the teacher, and for your school setting." This decision is not always a simple either-or situation. The prerogative for this decision may not always be yours. In some instances, you may inherit a program which you may be invited to improve. In others, you may be given a program with little or no allowance for your contributions. In still other instances, you may have complete responsibility for any and all of the science instruction that goes on in your classroom. All these instances involve some decisions. Do you try to live with what you inherited? Do you alter what you inherited? Do you accept, without struggle or input, the mandated program? Or, as the case may be, do you provide science instruction where none previously existed?

To meet any of these challenges, several courses of action may be considered. Science programs can be created, borrowed, or purchased; or, any combination of these three can be taken to meet the needs of your children. Currently, many excellent, so-called store boughten programs are available. Usually, the more inexperienced the science instructor the greater the need may be for well-structured, commercially prepared materials and the less the likelihood of self-created programs. Individuals who are new to the teaching profession and science instruction initially require more detailed, documented, supportive systems. These are available in the form of prepared texts, packaged materials, kits, and well-delineated teachers' guides. Often, such systems contain prepared lessons, dittos, and tests for evaluation. It would appear that a system like this, plus time and a commitment to the teaching of science, would be all an instructor needed to teach elementary school science. This has not always been the case. Excellent materials in the hands of inadequate science teachers may end up as inadequate programs of science instruction. And conversely, seemingly mediocre materials in the hands of creative, innovative teachers may become excellent programs. Whether you create, borrow, or purchase a science program, you first need to know what is currently available.

There are over 2400 colleges and universities in the United States. Almost all these institutions have one or more individuals who claim expertise in science or science education. Many of them are diligently investigating certain areas of science instruction. Their activities range from small-scale investigations of very discrete areas of science instruction to the development of full-scale projects spanning many years of the elementary curriculum. The results of all the activities of these individuals could not possibly be covered in this text.

The single most comprehensive source for finding out what is going on in science curricular development is the current *International Clearinghouse on Science and Mathematics Curricular Developments*. The University of Maryland's Science Teaching Center in College Park, Maryland, has taken on the task of summarizing and publishing the majority of ongoing projects in science and mathematics curricular development. This compendium of programs lists titles and subject areas covered and summarizes the goals and characteristics of each project as well as the evaluation techniques utilized. This publication contains approximately 400–500 pages and lists over 300 projects. Not all the projects would be directly applicable to your needs. Nevertheless, a perusal of them would give you a broad overview of where the science curricula has been, where it is now, and, perhaps, allow you to infer where it is going. Still, the question remains — Which program for me?

From the vast array of science instructional materials, we have selected three curriculum models: *Science — A Process Approach* (SAPA), *Science Curriculum Improvement Study* (SCIS), and *Elementary Science Study* (ESS).[4] These three have been selected because they represent complete programs for grades K–6. Two are based on the thinking of psychologists: SAPA, R. M. Gagné; SCIS, J. Piaget. Both are sequenced materials. All three programs challenge children to hypothesize, to collect and interpret data, to generalize, to create models, and to predict.

Science — A Process Approach (SAPA)

SAPA is an inquiry-oriented, elementary school science program based on the processes of science. Experiences that result in a cumulative and continually increasing degree of understanding and capability in the processes of science are stressed. These processes are divided into basic and integrated processes. The basic processes (observing, using space/time relationships, classifying, using numbers, measuring, communicating, predicting, and inferring) are assigned to the primary grade levels. The integrated processes (controlling variables, interpreting data, formulating hypotheses, defining operationally, and experimenting) are introduced at the intermediate grade levels.

The primary objective of Science — A Process Approach is to help children acquire competence in the processes of science. Much content is taught within the structure of SAPA (physics, chemistry, earth science, and biological science). There is more physics than chemistry in the primary-grade segments of the SAPA program, and more chemistry and earth science in the integrated intermediate-grade segments. Biological concepts form the bases for activities throughout all segments. Science content is not organized around a content hierarchy. Rather, the process skills are the hierarchical spine upon which content rests. Content development is subordinate to the development of science-process skills.

4. The publisher of SAPA is Ginn and Company; SCIS, Rand-McNally; and ESS, McGraw-Hill. Addresses and further information will be found in Appendix B.

This does not minimize content acquisition, for SAPA recognizes that there is no substitute for knowledge. Scientific processes cannot be taught without using content. Instruction in the processes of science is matched to the pertinent content to instill and enhance the processes. The SAPA program has seven levels, each with its own distribution of content and processes.

Science Curriculum Improvement Study (SCIS)

The SCIS program is concerned with developing children's thinking processes — from concrete to abstract — by disciplining children's curiosity. Appropriate investigative experiences in biological and physical science coupled with a maintained and disciplined curiosity fosters scientific literacy. Paramount to the SCIS program is the notion that changes take place because objects interact in reproducible ways under similar conditions. And, the SCIS program capitalizes on four major scientific concepts to elaborate the interaction concept — matter, energy, organism, and ecosystem. In contrast to SAPA, where the content of science is subordinate to the processes of science, SCIS stresses content more than processes.

Children are provided materials with which they are allowed to interact in their own way. They explore. They invent concepts from analyses of the data which are the results of their explorations. They discover information about the subject which is being investigated.

Transmittal of information is not the primary purpose of the SCIS program. Rather, the primary purpose is to educate children to "search out" information on their own in a disciplined manner. The SCIS program comprises seven branches. These seven levels follow two main stems:

	Physical science units	Life science units
Kindergarten	Beginnings	
First level	Material objects	Organisms
Second level	Interaction and systems	Life cycles
Third level	Subsystems and variables	Populations
Fourth level	Relative position and motion	Environments
Fifth level	Energy sources	Communities
Sixth level	Models: Electric and magnetic interaction	Ecosystems

The SCIS program organizes the content of science to lead to specific conceptual structures about matter, energy, organisms, and ecosystems. Exploration, invention, and discovery lead children to find these structures. The process skills of science are imbedded in the acquisition of content but not independently taught.

Elementary Science Study (ESS)

The ESS program is as unlike a program as any program can be. Its origin is not rooted in any concept of how science is organized or any theories of how it should be taught. Rather, it is anchored in sound scientific activities that children actually enjoy. All materials developed by ESS were subjected to rigorous classroom testing under a number of conditions and with a variety of children. Based on the interest exhibited by children and teachers, these activities were organized into units. These units were published along with teacher's guides, films, film loops, kits, and printed materials. The rationale for this back-to-science instruction is based on the ESS philosophy that no group can design a single science curriculum that will be all things to all people. Therefore, the program provides units from which teachers can select to organize their own curriculum. To assist a school in preparing its own curriculum, ESS has prepared a printout with a recommended grade range for each science unit. Teachers can assemble a balance of science topics from the various science disciplines using the ESS unit list.

The ESS units tend to be more open-ended than most prepared materials. The units are most effective in classrooms where inquiry is encouraged, where children are invited to be more independent, where teachers talk less and listen more, and where structured learning is minimized.

Although the table is an oversimplification, it summarizes important features of the three curriculum models.

The decision is yours

You may still be asking the question: Which program is best? All three of these curriculum models are excellent. In addition, many other fine, commercially prepared science texts and materials are available. Some of these newer programs have their foundations in one or more of the three curriculum models.

To repeat, there is no shortage of useful materials in science instruction. The best program is the one which you judge best for your students

	Grade level	Psycho-logical base	Organiza-tional stress	Science emphasis	Inquiry oriented?
SAPA	K–6	Gagné	Process over content	No specific science discipline empha-sized or sequenced	yes
SCIS	K–6	Piaget	Content over process	Physical and bio-logical disciplines empha-sized and sequenced	yes
ESS	K–8	—	Open-ended; single unit presented	No specific science discipline empha-sized or sequenced	yes

and that you feel most comfortable teaching. Some school systems use one or any combination of these programs. Some school systems use none. Your responsibility is to be knowledgeable about existing programs and cognizant of where and how you can improve your science curriculum. It is not necessary to adopt programs but to adapt them for maximum learning in science for the children in your classroom.

WHY BOTHER?

Even though one might accept science as a "many splendor'd thing" and worth teaching, it would be easy to excuse oneself from this task. Yet many accept the challenge, in spite of obstacles that might have deterred them. Their positive action stems from the acceptance of certain basic tenets. A few of these tenets are that each student is capable of learning the basic principles and concepts of science; that each child can experience joy in the recognition and acquisition of the intellectual powers of science; and the realization that this is it — for many students, science instruction terminates at either the eighth or ninth grade.

A contradiction appears to exist between the attributes, inferred or otherwise, of science and its position in the order of priorities of teachable topics. Teachers offer many reasons for the stated priorities of subjects. Many are valid. Rather than argue the wisdom of such decisions, though, let's look at the conditions that exist. Some negative statements offered by teachers for not teaching science or for teaching it less frequently are accompanied here by positive responses:

Negative statement: There isn't enough time in the day for all the topics. First things first. Reading, language art, and math are first.

Positive response: All fields are inextricably related by fundamental principles and concepts; divisions between them are artificial, erect barriers to the integration of the fields, and indicate man's inadequacies.

One cannot really teach *one* subject. One of the processes of science is communication. Communication is a process not only of science but of all human endeavors. Scientists communicate with oral and written words, diagrams, maps, graphs, mathematical equations, and various kinds of visual demonstrations. Scientists read and readers often become scientists.

Another direction supporting the integration of the various topics is the trend away from science that is built upon a limited understanding of mathematics toward science built upon a good grasp of mathematics.

Negative statement: Certain skills such as reading, language arts, and mathematics are prerequisite for success in school and must be covered.

Positive response: The central purpose of education as stated by the Educational Policies Commission is: "The purpose which runs through and strengthens all other educational purposes — the common thread of education — is the development of the ability to think." [5]

Evidence increases that experiences in creative sciencing teach

5. Educational Policies Commission, *The Central Purpose of American Education* (Washington, D.C.: National Education Association, 1961), p. 12.

children to think. A recent study indicates that new approaches to science are superior because they:

1. help children to develop reading-readiness skills;
2. develop the process skills of science better than textbooks;
3. promote scientific literacy and intellectual development;
4. help children to apply mathematics in problem-solving situations;
5. help children in the curricular area of social studies skills including interpretation of graphs, tables, and posters and the reading of maps.[6]

In addition to being good science and developing children's ability to think, creative-sciencing approaches also make for good reading, mathematics, social studies, and art. A further benefit seems to be that these approaches help children become creative, innovative, and independent decision makers.

Negative statement: I don't teach science because I am not adept in science. I don't feel comfortable teaching it. I have an extremely weak background in science.

Positive response: Some curriculum trends that have helped teachers to feel more comfortable in teaching science are the trend away from much subject matter to less; the emphasis in science away from accumulating facts to an emphasis on how to find and generate knowledge through inquiry; the movement away from teacher-selected concepts as teaching goals to concepts as they arise in confirming or rejecting hypotheses; and a move from science learned wholly from books to something that grows out of a series of experiments.

Whereas teachers were once reluctant to teach science unless they were proficient in it, now more teachers are willing to engage in science and learn with the students. "I don't know" is now as common a teacher's expression as "Let's find out together." Acknowledging there is no substitute for knowledge, we must also point out that no person can possibly know all there is to know about science, or anything else, for that matter. The trend toward focusing on skill in inquiry rather than facts and factual concepts has made the teach-

6. John W. Renner, Don G. Stafford, William J. Coffia, Donald H. Kellogg, and M. C. Weber, "An Evaluation of the Science Curriculum Improvement Study," *School Science and Mathematics,* vol. 73, no. 4 (April 1973), pp. 291–318.

ing of science a less frightening experience. The spirit of open-ended and divergent approaches to problems in science has created a more receptive climate for teachers who feel somewhat inadequate relative to the content of science.

The acceptance of instruction in the processes of science — methods, techniques, strategies, and tactics as being far more important in teaching science than merely its product — has helped place teachers at ease and has increased science instruction in schools.

Negative statement: Science is messy. It is a lot like teaching art; it takes a lot of paraphernalia. The kids get into everything. It isn't tidy. Science takes too much equipment. It is jars, beakers, thermometers, hot plates. I don't have any place to store this equipment if I had it. And, if I had it, I don't have any real working area where I can use it.

Positive response: No doubt about it. Teaching science through inquiry and involving students in a hands-on type of instruction requires equipment. Surprisingly enough, it takes little elaborate or expensive equipment. For too long teachers have withheld science from the curriculum or relegated it to demonstrations because of limitations in equipment. Some items are fundamental. You may need a minimum amount of Pyrex glassware (test tubes, beakers, and the like). You may need a timer. Uniform dripping tap water, a beating pulse rate, and so on, are all right, but you will need an accurate timepiece. Magnifying glasses are important. So are balances. Adequate substitutes for these can be constructed or improvised.

The catalyst in creative sciencing is what you can envision as part of your science program when you view various articles. This is like being a pack rat with purpose. Most teachers can recognize that the plastic covers on aerosol cans can be used for something; that empty ketchup bottles can be useful; or that aluminum trays can be valuable to the ongoing science program. It takes dedication to alert oneself and one's friends to save various items such as graduated peanut butter jars, cigar boxes, or large mayonnaise jars. No teacher, on instant demand, can quickly assemble thirty baby bottles, thirty tuna fish cans, or thirty pie plates. And yet, all these may be needed to involve the entire class in a science experiment. These items must be gathered in advance over a period of time and stored (garages, attics, closets, storage boxes, almost any place will do). Teachers (or children) who collect, sort, classify, and catalogue materials find that this technique of providing enough equipment, materials, and supplies for everyone is not a formidable task and moves the science program forward.

PAUSE FOR A SUMMARY

— What we say and what we do as teachers can have a lasting effect upon children.

— Creative sciencing advocates the integration of all areas of the curriculum with the attitudes, ideals, and spirit of science.

— Science creatively develops observational skills; skills in making distinctions, in critical thinking, and in experimenting.

— The basic purpose of education is to help students learn to think.

— Science is a natural for developing attitudes, thinking skills, and motor skills.

— Determination of the content in the elementary science curriculum depends on several factors including societal goals, the structure of the sciences, and learner goals.

— Elementary school science should be exploratory in nature with the learner, rather than the structure of the sciences, dictating the curriculum.

— The intermediate and middle school curriculum should center on expanding science to establish the continuity among its disciplines.

— There is no shortage of outstanding materials available for good science instruction in the elementary school.

Creative sciencing
–creative teaching

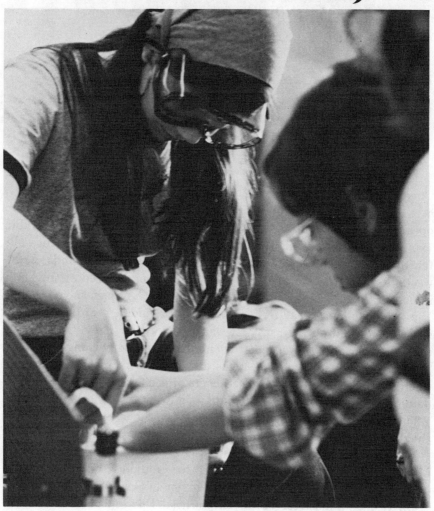

If it's too much trouble,
it's probably worth it!

Where children are encouraged to make many choices, they increase in confidence and self-awareness. . . . If children have many choices to make, the teacher is able to observe each child in a wide variety of situations. The broader her observation, the better she is able to plan for a child's further learning.
— Laura D. Dittmann

A TRUE FAIRY TALE

Once upon a time long ago (it really only happened a minute ago) a new, open-concept, continuous-progress, individualized, self-pacing school opened. It was a beautiful sight — the interior was shiny and sparkling (you could dine on the polished tile floor). There were intermittent walls, few doors, no room designations, and nothing was out of bounds. As the week preceding the start of school began, teachers arrived at the building and readied themselves for the new year. They met their principal (he had a one-hour break from his role as tour guide), heard his inspiring speech, and went on their way. It was a busy week. They had to set up the bulletin boards to welcome the children to school, arrange the desks in neat rows, divide the coat racks and storage cabinets, and arrange them to encircle each area. The teachers of each grade level had to predetermine the division of the children into groups (the blues, greens, golds, and purples), order the movies, prepare the lesson plans, and post the rules for conduct in the school.

Happy children with smiling faces entered the school the following Monday morning. They were pretested for grouping in reading, mathematics, science, and social studies (the major subjects).

Oh yes, we almost forgot to tell you that this school used the integrated-day approach, except:

Art *had to be* from 9:00 to 9:50 on Mondays and Wednesdays.

Music *had to be* from 1:00 to 2:15 on Thursdays.

Physical Education *had to be* from 2:00 to 2:30 on Fridays, and so on.

At the end of the first week the principal and his administrative staff (the secretary, the coach, and the janitor) decided that homerooms should be instituted, based upon grade level.

At the end of the second week the teachers requested that the children remain with them all morning for skill work in a self-contained setting.

At the end of the third week the skill work was continued all day.

By the second semester dividers had been installed throughout the school along with room doors. The children stayed with their one teacher all day, except for "specials."

And they all lived happily ever after; all, except for the children.

Educational jargon is overwhelming. Some of the jargon words are used in our *true* fairy tale. Unfortunately, in education, more often than not, the vocabulary changes but the instruction doesn't. Before we present you with skills and techniques that are useful as alternatives to the traditional instruction, let us examine alternative approaches to traditional educational problems and the use of creative-sciencing aids.

CHANGE FOR CHILDREN

Things must change so that they can remain the same. Change is universal. Change is life. You are not the same at this instant as you were when you started to read this paragraph. Education has changed. Education must change. The change should not be for change's sake, but it should be in response to educational needs as they arise.

Creative-sciencing teachers are not afraid to change. The words, "I tried that and it didn't work, so I'm trying another plan of attack today," are often used. The creative-sciencing teachers' motto is, "Change for children." No idea, method, or activity that has potential for helping children to learn and to succeed is eliminated before it has been tested and evaluated in the classroom.

Creative-sciencing teachers place children first. The children are their overriding concern, more important than the school, the superintendent, the school board, other teachers, jobs, or anything else.

Think of how you would change to meet the following situations:

— if children came to school only when they wanted to;

— if you could teach what you wanted to, when you wanted to, as long as you wanted to;

— if you had no supervisors;

— if you got paid by the number of students you attracted to your class;

— if you could regulate the curriculum;

— if you did not have to worry about the calendar;

— if you got paid in direct proportion to the scholastic achievements of your students;

— if teachers were not granted tenure;

— if you taught eleven months of the year;

— if children chose their teachers and changed teachers as they saw fit;

— if you were asked to televise your lessons to children being taught at home several days a week instead of attending school five days a week;

— if children graded teachers.

Problems are more plentiful than solutions. It is easy to raise questions. It is easier to give advice. Were it not so easy, not everyone would be so eager to donate something to this pastime. Solutions to problems are bandied about by professionals who expound and then resolve the problem by using the eternal, "We ought to. . . ." When the din fades away and you reflect on the *we*, you will quickly find that *we* is *you*. Authorities can identify problems, society can demand satisfactory solutions, but final solutions rest in the hands of classroom teachers. The individual teacher is the vital catalyst in the process of education. The *we* must be *you* or woe is *us*.

This segment of *Creative Sciencing* explores areas to expand your role as an agent of change.

CHANGING YOUR SCHOOL

Grandmother, daughter, and granddaughter were all reminiscing about their school days. Each in turn described a typical classroom of her era. At the finish they looked at each other in amazement. Each had described the same room. Well, not really, but nearly so. So few changes in classrooms had taken place in the sixty-five-year gap between grandmother's and granddaughter's time that, for all practical purposes, they seemed to be talking about the same room. Is your classroom much the same as the classrooms of your childhood? Most assuredly paint colors and furniture have changed, but it is surprising that classrooms now are more like than unlike classrooms of the past. What better place is there to start than with one's own classroom and oneself when changing the school?

CHANGING YOUR PHILOSOPHY

Every teacher has a philosophy underlying classroom teaching, whether it is written or unwritten, spoken or unspoken. Our actions reveal our philosophy. We can hardly conceal our views of education. Our every action bespeaks our beliefs. Some of us are steadfast in our philosophies; others are meandering, searching for a philosophy. In either event, review these thoughts — some may cause you to alter your philosophy.

As you teach children the many wondrous things of life, remember you too are growing, changing, maturing, learning, loving, just like your students. You and they are eternal learners.

Education is not just preparation for life, it *is* a portion of life. Education is a highly personal, fragile thing. It is nurtured internally, digested slowly, and is powerful enough to move the universe. It is a glorious thing you do, encouraging learning in others.

Take some risk. A turtle can only progress when his neck is stuck out. Push a little for what you believe in.

Start each day thinking — to children with love.

Raise your sights. Aim high. Even if you miss the highest target, you may strike a target of value a little lower down. If you aim low and miss, you will strike the dust. Children should leave your class one step closer to being creative, innovative, independent thinkers. Aim high.

Being a good teacher is like being a good parent — at times, extremely difficult. At times you will not know if what you are doing is right or wrong until it is too late to change anything about it. But nevertheless, you must make decisions.

You don't have to know everything, but it helps. Improve your posture. Be yourself. Relax and let the good teaching roll out. Think positively. You are OK. The children are great. The sun rises every day even though some days are cloudy. Remember you may not be the best teacher in the world, but you'll do until he or she comes along.

Changing Your Philosophy / 31

Slow down. Most teachers have a tendency to rush learning. A teacher cannot teach faster than the students can learn. Allow children time to reflect on and respond to your teaching and their learning.

CHANGING YOUR ATTITUDE

Children are amazing. They are basically patient with, tolerant of, and sympathetic to the inconsistencies of education. Some even love us. The loyalty that children exhibit toward the system is a reflection of their acceptance, or perhaps resignation, during 180 days a year for at least nine years. In any event they are deserving of a round of applause. This applause should be tempered with awareness that not everyone you teach will have the same excitement for learning that you have. Different appetites want different menus. Avoid the negatives in teaching. *Don't, can't, won't, haven't,* and similar words should be supplanted, where possible, by positive statements. Confidence is a good teacher trait. It is also a good child trait.

Love children. If you can't love them at least like them. Not all children are easy to like but, nevertheless, they welcome and need affection.

Smile. Don't wait until Thanksgiving Day to launch your personality. Laugh. It is amazing how few students can describe their teachers' laughter. It happens too rarely. Laugh with the children not at them.

Rarely do teachers fail because they are not intelligent enough to teach. Most fail because they lack the personality to communicate and relate to children and adults. No one owes you anything. You owe yourself something. Smile. Enjoy the job. Teaching is fun. Children are great. Give a little — then a lot.

gold and silver have their price —
learning is priceless

CHANGING YOUR OBJECTIVES

In preparing to teach a specific lesson a teacher might legitimately ask two questions. *What do I want the students to learn?* A response to this question necessitates a rationale for the instruction and performance objectives (or what the students will be doing to acquire the learning).

Do the students already know this? A response to this question may be found through preassessment measures.

A major complaint about performance objectives is that teachers tend to prepare the majority of their objectives along the path of least resistance. This path follows the cognitive, or knowledge, domain. Performance objectives that merit consideration and yet receive less attention are those concerned with the development of attitudes, values, interests, and appreciations — all part of the affective domain. The objectives of the psychomotor domain are also slighted in favor of objectives in the cognitive domain. Teachers are rightly faulted for restricting performance objectives in the cognitive domain to simple recall rather than spreading them throughout the hierarchy of thinking skills. Little attention is directed to higher orders of thinking such as analysis and synthesis.

Change your objectives by including a variety of levels within each domain, be it the cognitive, affective, or psychomotor.[1] When preparing performance objectives, you may wish to consider the fact that we don't demand accuracy in art or creative writing, but we have permitted ourselves to require accuracy in science. We may be paying a high price in lost interest, enthusiasm, vitality, and creativity in science because of this requirement of accuracy. Science is not always exact. It is a creative engagement with ideas to solve problems. Many proposed ideas, at first, are seemingly harebrained, but they may well give rise to a brilliant solution.

CHANGING THE CURRICULUM

Adults make 99.9 percent of the decisions relative to what children learn, when they learn it, and how they learn it. You might improve your curriculum if you believe that children have the capacity:

— to make decisions when confronted with decision-making situations;
— to assume responsibility in direct proportion to the part they play

1. For more information, see Benjamin S. Bloom ed., *Taxonomy of Educational Objectives, Handbook I: Cognitive Domain* (New York: McKay, 1956); David R. Krathwohl et al., *Taxonomy of Educational Objectives, Handbook II: Affective Domain* (New York: McKay, 1964); and Anita J. Harrow, *A Taxonomy of the Psychomotor Domain: A Guide for Developing Behavioral Objectives* (New York: McKay, 1972).

in delineating and assigning responsibility (for children become dependable when they are depended upon);

— to learn at their own pace;

— to like what they must do as well as to do what they like;

— to improve in decision making when they have experience as decision makers;

— to respect themselves as well as others;

— to learn from each other (for interaction with peers is important to the child's total learning process);

— to learn best when they can see the relevance of the learning;

— to want to succeed and to please you, parents, and themselves.

Set up a periodic hour for "druthers." What would they "druther" do than ... ? Then do it together.

REMOVING THE EDUCATIONAL STRANGLEHOLD

A favorite story in educational circles tells of a time in the late 1960s when a European educator visiting the United States chose to investigate a so-called modern school. On visiting a fair-sized community and meeting a fair-sized superintendent, he inquired, "What is a modern school?" The superintendent hesitated momentarily and then said: "I don't understand it myself. I know only two people who understand it. One is a second grade elementary teacher on the east side of town and the other is one of our custodians in this building. Unfortunately, these two don't agree."

Things are not that bad, but sometimes they approach that level of confusion. We talk of traditional schools, modern schools, open-concept schools, and so on, ad infinitum. To refer to any one type as unique is, of course, totally unwarranted and incorrect. Each has a bit of the other. Regardless of the title, each has a bit of a stranglehold on children. The title may be different, but the noose is the same.

Nooses are formed by dedicated teachers who have spent some sixteen or seventeen years preparing to teach, and with the help of the Almighty, they intend to teach. And teach they do. Being intelligent, red-blooded, strongly motivated, well organized, hardy in health, and determined to succeed, they plunge forward attacking the besieged learners with a barrage of facts, concepts, generalizations, all of which are supposed to result in significant learning for students. The end result may be summed up thus: "Never have so many waited so long for so little." This commitment to exposition has existed long enough for researchers to have evolved what is termed *the rule of two-thirds*. The rule of two-thirds says that, on the average, in elementary and secondary schools, almost irrespective of subject areas, two-thirds of the class time is spent talking — two-thirds of that time the teacher is talking, and two-thirds of the teacher's talk is telling or demonstrating rather than interacting with students.

Many teachers seem compelled to see to it that every detail of the knowledge they feel responsible to transmit is transmitted. It must be crammed into the hapless and often numb children. But children are nice — they still love us. Well, at least they tolerate us, despite our attempt to strangle them with knowledge.

It doesn't matter what setting you operate in. It doesn't matter what title you assign to your teaching strategy. It does matter that you create a learning environment that will recognize and respond to individual needs and differences.

This commitment carries with it a prime consideration to change from a pedantic dispenser of knowledge to a facilitator of learning and a responder to children's behavior. Such a change results in *inquiry learning*. The important location is no longer in front of your audience, it is among your audience. No longer do teachers perform in front of children; they set the stage and then move out of the way so learning by children can take place.

The age-old question, Who learns the most in a classroom?, need not always be answered: "The teacher, naturally!" This may seem a valid response inasmuch as some teachers do most of the work, most of the thinking, all the talking, all the questioning, and in many cases most of the answering. We know this approach impedes individual progress. It shortchanges the learner by making him or her a spectator rather than a participator in learning.

Only to the extent that children make real choices in identifying their goals (planning strategies for learning and evaluating their progress by questioning themselves and others) can real learning take place.

We may never know exactly how we learn or how any other individual learns. Students learn by themselves as well as with their peers and teachers. It is said that no man is an island unto himself, but in some ways he is. Learning is personal, internal, and unique to the individual. Each child learns for himself or herself. We have done a poor job of convincing children that learning is for them. Most children go to school to please parents, teachers, and truant officers.

There is a joy to learning. There is an intellectual power to learning. If there are elements of belief and trust in a learning environment, if the value of each child is respected, and if each child is free from total dependence on a particular individual or particular approach to learning, perhaps this joy and power can be realized.

CHANGING YOUR
QUESTIONING TECHNIQUE

Teachers talk. They talk a lot. Often they talk too much.

Teachers ask questions. They ask a lot of questions. Often they ask too many questions.

Recent tallies of teacher-pupil interaction revealed that 60 to 70 percent of the words spoken in a classroom were spoken by the teacher. The remaining 30 to 40 percent of the spoken words divided by the class population reduces down to a small number of spoken words per student, per day. On the average, for every question asked by a pupil, a teacher asked twenty-seven questions. The teacher's questioning rate averaged three-and-one-half questions a minute. Unfortunately, this domination does not develop an aggressive attitude toward thinking in students. Most questions teachers ask are trivial. Few stimulate students to think, and few draw well-thought-out responses from students. Most teachers' questions call for answers falling into the categories of memory, information giving, criticism, and comparison, with the heavy emphasis on memory and information giving. Rarely are questions directed to a problem-solving situation. The teachers' approach to questioning is reflected in the poor questions asked by students. Students' questions usually reveal little interest or serious thought. When teachers ask, "Are there any questions?" the response is meek and weak.

It is clear that in the teaching-learning situation, the teacher is truly the dominant figure. Teachers are prone to try to do students' thinking instead of motivating the students to think for themselves. Teachers who dominate the learning scene do little to stimulate inquiry or creativity. In fact, they discourage it. Inasmuch as so many of the questions teachers direct at students require only the use of memory to answer them, students tend to think the main purpose of education is memorization.

Teachers concerned with training students who engage in the pursuit of learning for the purpose of understanding trade the *quantity* of questions for the *quality* of questions.

Questioning is akin to good science investigating. A scientist formulating a well-thought-out hypothesis is questioning. If one asks good questions, one usually gets good answers. Teachers who practice poor questioning techniques promote student guessing and slovenly habits of thinking. Students, as well as teachers, should be aware of good questioning techniques. The teacher needs to provide the model.

Most successful teachers are good questioners. They ask open-ended questions that promote discussion rather than closed questions that call for single-word answers.

Open-ended questions	Closed questions
What do you observe?	What color is it?
Can you suggest a way to classify these?	Can you classify this by color?
What inferences can you make?	Is the cloth wet?

Often closed questions are strictly of the yes-no type, and they evoke little discussion or student involvement. Open-ended questions promote discussion and often require the student to use decision-making skills.

Creative-sciencing teachers promote discussion by asking the question *first* and then calling a student by name to answer it. They may ask another student to comment, and another, then another, rather than simply respond to one student's answer.

remember:

Creative-sciencing teachers encourage student-to-student interaction rather than discourage it.

Creative-sciencing teachers allow sufficient *time* for student responses to questions. They practice waiting from five to fifteen seconds for each student response rather than using the machine gun approach — a question every two or three seconds. Open-ended questions cannot be answered in one or two seconds. Children need time to respond to questions and *they need to have their answers accepted*, temporarily at least.

How many times will a student try to respond to a question if continually told that his or her answers are wrong? Wrong answers should be reserved for quiz shows on television. Children should be encouraged to respond to questions; and their answers should be accepted until a "better" answer is discovered either through discussion, experimentation, or in reference materials.

Creative-sciencing teachers are not afraid to say, "I'm not sure," "I don't know," "Maybe you're right and I'm wrong," "That's a great idea." They think positively and encourage lots of happy faces rather than sad ones.

Creative-sciencing teachers have developed many positive responses to student answers rather than just OK, yeah, right, or wrong. They include:

Good job!
Well done!
Good for you!
Wow!!
I like that!
You're doing a great job!
Outstanding!
Very nice!
Superior!
I'm impressed.
Interesting.
Excellent!
Good!
Super!

Creative-sciencing teachers are concerned about their students' feelings and often ask them to evaluate what has been occurring in their science class. Is science their favorite subject? Why or why not?

Are they doing things that they don't like? How could this be changed?

What do they like or prefer?

What would they like to change?

Evaluation is a continual process, *not* a once-a-semester or once-a-year thing.

Try improving your questioning technique by:

— limiting yourself to one question in a fifteen-minute period;
— never answering your own questions;
— eliminating questions that call for a yes-or-no response;
— not asking questions that invite aimless or guessing responses;
— asking each child in your class at least one meaningful question a day;
— asking both narrow- and broad-response questions;
— limiting memory questions to one an hour;
— letting the students ask the questions;
— mixing up your questions (thought-provoking questions, what-if questions, evaluation questions, and a variety of other types);
— asking a question and then listening. (Teachers are notoriously poor listeners, particularly with children.)

CHANGING YOUR DISCUSSIONS

Talk to students. Don't always talk teaching-learning talk. Children need the faith and courage that come from your consideration of them as individuals who can make contributions. They need this more than they need to know the associative or distributive law or how to spell the word *artichoke*. Talk to them about many things. Talk to them as you would talk to invited guests in your home. If necessary, change your conversational style. Communicate with individual children, not classes of children. Make every child feel that you are talking directly to and for him or her.

Listen to students. Consider all students' responses. They say many things that have implications for improving your teaching. Learning that significantly influences behavior is self-discovered, self-appropriated learning. Instead of directing, ordering, meting out assignments, monitoring a practical rest-room schedule, or disciplining, step off the merry-go-round and strike out in new directions.

You say you can't. Remember the old adages pronounced by elderly wise men or women, like: "Where there's a will there's a way." "If you want to do something badly enough, you'll do it." "If there is something you want to do badly enough, you'll find the time." By and large these statements are true. We are not advocating a complete remodeling of your teaching style but we are advocating consideration by you of the changing role of a teacher. Your discussions should reflect this changing role. You are not a dispenser of knowledge but rather one who elicits and clarifies learning. You are the one who convinces children that they have worth and that you have faith in them. You are the mood or climate setter. You are an organizer and facilitator of learning. And you recognize that you have limitations.

CHANGING HARD ROOMS TO SOFT ROOMS

Most children look forward to their first year of school with indescribable anticipation. They love their teachers. A book bag and a pencil box are all an exciting part of the grown-up process of getting educated. This enthusiasm slowly dims until it bottoms out in what is called the fourth-grade slump. The excitement has worn off. September school attendance is only highlighted as a replacement for a long summer vacation. After the first few days when old friends have been revisited and new teachers scrutinized, rationalized, and categorized, the resignation sets in. It never ceases to be a wonder that so many students can tolerate so much so long with such reasonable — under the circumstances — politeness.

The decline of interest in school can be attributed largely to the artificial environment that exists in schools. It is unnatural for students to become a captive audience corralled in a room or several rooms and subjected to teachers who have decided in advance who shall learn what, when, and to what extent, and in what given period of time.

The classroom should be an extension of the home. Granted, children in your class may come from a variety of homes — some far better than any classroom, some as good, and some grossly inferior to any classroom. This comparison is made in terms of furniture, fixtures, heating — the list is endless. In terms of what individuals need, any school, the best school, is inferior because the school exists for adults not children. Children rarely think of the school as their home although they are invited to do so by principals. And yet, many children spend more waking hours in school than in any other place.

What does a school lack, or have, that causes students to reject it as their home? The biggest single factor is that there is no place to hide. This factor affects teachers as well as students. There is no place for privacy. There is no place for private thoughts.

Not only is privacy lacking, but everything needs a sanction. At home you don't need permission to invade the refrigerator, retreat to the basement, use the bathroom, or step out-of-doors. In school you can't talk,

LISTEN TO THE SOUNDS
A CHILD MAKES

by Patricia R. Burgette

Listen to the sounds a child makes,
The spoken pleas, the silent message,
It is the child we teach,
Not the content, the morals, the skills.

Listen for the child who needs to be seen.
He needs to be known as unique,
Not just another one of the group.
The message may come
As a flurry of temper,
A shouted word,
A carefully spoken question,
Or a soft intent look.

Listen to our own needs,
The problems of the teacher,
The principal, the counselor.
How are these needs met?
Whose desires determine the course of the day?
All are there and should be heard,
Never forgotten or put aside.
We are all there and all must be a part.

Listen to the world outside,
Beyond the walls of the room.
Move as a part of life;
Share the warmth and reality of today.

can't laugh, can't move, can't shout hallelujah, can't whistle. You sit or you move in a prescribed manner. You play games. You have to please one adult continuously.

The parameters for running a school rule out its ever becoming the equivalent of a home. This need not be. Children know this. Teachers know this. Principals know this. Schools have other things to offer that a home could never have. Schools have, for example, thirty or more constant companions, a team, a spirit, a publication, and so on. We all know what is desirable but many of us are reluctant to teach as well as we know how. We teach as we have been taught. We honor our past heritage even though we know better. Somehow we feel this is the way we are expected to act.

Traditional, safe teaching is hastily saluted and the merit raises go on. Teaching and reaching out are a bit more frightening. The ground is not as solid as in a subject area. Statements like, "If I let the children do as they please, it will be chaos, they'll never learn; they won't be ready," show a few of the fears of teachers making a transition to an alternative approach in the classroom. What does one do to soften up a classroom? Here are a few suggestions:

— Periodically rearrange the room. Let the students assist you in deciding what to place where.
— Arrange the room so students can have private areas.
— Bring in a rocking chair. A recliner would do nicely. Why not! Some people study best in a prone position, others in a rocker.
— Have a few radios available. Periodic news and weather forecasts are part of the curriculum of living.
— Have a junk box or "creativity bin" available.
— Provide greater storage space for children. The only private areas most children have are their desks. And those aren't very private.
— Allow children to get involved in decorating the room.
— Allow their ideas to come through. Remember it's their room too.
— Eliminate requests for permissions. Have the class provide the ground rules.
— Let each child get involved in a project of his choosing.
— Bring in an old refrigerator. Let the class rule on what the behavior will be.
— Look at every conceivable space and surface area in light of a new purpose. What can you do with a ceiling? Is a floor just a floor? Are you really short of bulletin board space?

Creative Sciencing — Creative Teaching / 42

— How many musical instruments do you have readily available in the room?

— If your room doesn't already have a rug, get a rug. Even a small one will do.

— Ask individuals to describe the conditions under which they work best. Try to duplicate some of the conditions that suit the largest number of students.

Some critics would howl long and hard at these changes. Some teachers might renege on accepting any or all of these changes. Some principals would be upset if the school looked too "lived in," if the children looked as though they were loafing instead of learning, or if the noise were beyond their personal expectations. They would be right to criticize. Anything can be carried to extremes. Anything that resembles a free-swinging circus with no learning going on can be an educational disaster. The rub is — how does one know when real learning is going on? There is no assurance that any more, or any less, learning is taking place in a "traditional" as opposed to a "modern" classroom. The only consolation a teacher has is in the acceptance of the notion that learning occurs best when individuals have a greater degree of freedom to explore, manipulate, and experiment within an environment. Associated with this latitude must be the provision of conditions that foster an attitude of "search" for learning. This attitude must be cultivated in an atmosphere of acceptance and approval, but at times it must forbid specific acts by individuals.

The learning process is to teach students to exercise individual control — control of themselves. If rigid controls are forever clamped on students, when do they mature?

A teacher doesn't need to be an educational interior decorator to make a sterile room come alive. Revitalizing a classroom can be a pleasant challenge, if you think of yourself as a merchandiser. You are in business — the business of education. You are selling a product. To make your room livable, with learning space that entices learning, try adapting some standard business practices.

— Change the display counter. (Alter the decor, bulletin boards, interest centers, and anything else to improve the teaching environment.)

— Issue credit. (Give everyone an A at the start of the semester.)

— Let the customers test-drive the product. (Provide some of the benefits that supposedly accrue from education.)

— Give refunds. (Anyone unhappy with his education can come back for a refund.)
— Exchange the product when the customer is dissatisfied. (Remember there should be no unhappy customers.)
— Offer discounts, bonuses, stamps, and other incentives.
— Have a clearance sale. (Raffle off incompletes, Ds and Fs.)
— Establish a preferred mailing list to alert interested students to unique learning events.

Some children's suggestions have evolved into highly satisfactory re-habilitations of so-called traditional classrooms. Plans for two such re-habilitations follow:

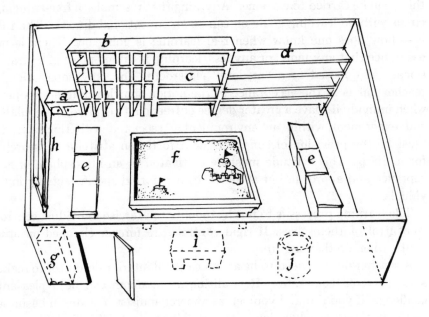

Room 1   The components, and their uses, are:

a. storage of large objects such as maps, large posters and so on
b. individual cubbyholes for each student, for the purpose of storing projects
c. storage of equipment needed in an action-involvement classroom, such as microscopes, balances, cameras, flashlights, and globes
d. a magazine rack for periodicals and books

e. students' desks or tables

f. a raised, off-the-floor, center-of-the-classroom sandbox

g. collection and specimen cabinets

h. blackboard and movie screen

i. teacher's desk

j. creativity bin for odds and ends

The most unusual feature of this room is the center-of-the-room, raised sandbox, with the height suitable for the grade level for which it is being used. This sandbox does what the blackboard cannot do — it gives you the third dimension. Using a blackboard conveys the impression that we live in a two-dimensional world. How much easier it would be to reconstruct the Battle of Gettysburg by quickly scooping out areas, building up others, inserting a road, adding a farmhouse, establishing a railroad, and developing other details, than to read about it. The battle can be reviewed as you develop and change the conditions within the sandbox. What better way to teach history and geography?

Also what better way to teach the volume of a cone and cylinder? Using a cone and a cylinder of similar heights, the volume of a sand-filled cone when poured into the cylinder fills the cylinder one-third full. Seeing, feeling, and listening make learning more lasting.

Mapping exercises, geology units, social studies units, reenacting reading stories — your imagination is the only limit — all done in relation to the sandbox concept, can enhance learning. What else could you add or change in this room to improve instruction?

Room 2 Here we have another classroom designed by children. The components are:

a. blackboard and movie screen

b. storage and display space

c. teacher's desk

d. students' desks or tables

e. book and reading shelves

f. mobile multipurpose cart

g. creativity bin

h. storage of sporting equipment

i. storage space for radio, hi-fi equipment, records

j. sink

k. storage space

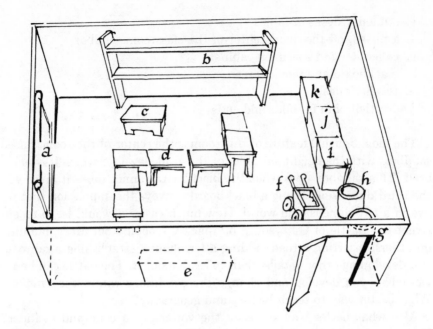

The primary asset of this room is its flexibility. The teacher is in the audience. The students are free to move and work in a place of their choice. The mobility of the furniture and equipment permits arrangements to meet a variety of learning situations.

In designing a creative classroom, think of the spaces that you may not currently use such as the window space, the ceiling, the wall space outside your classroom, and other spaces. Here are some suggestions: Think of the ceiling as the sky. Several constellations can be shown. The path of the sun and the phases of the moon may be plotted. Somewhere in the room note true north and magnetic north. Other compass directions should also be recorded around the room.

Windows have many uses. They can be used for decorative purposes or for experiments with light and temperature changes. Beekeeping is a fascinating window observation exercise. Your window, as the inside wall of the hive, can make everything easily visible. One teacher built birdhouses and feeders so that things could be viewed from inside the room. Herb gardens are also great window activities.

Hallways are great for displays. If you need to string a wire across the room to support draperies, fishnet artistry, or hanging gardens, do it! Just take the necessary precautions to insure that what goes up will stay up until you want it to come down.

Many valuable and highly usable room items such as display cases, advertisements, or card display racks are yours just for the asking. When you spot something in some market or shopping center that looks usable, ask: "What happens to that when you (the grocer, the druggist, the lumberyard man) want to get rid of it? Really! Please save it for me. Call me collect! The number is. . . ."

A local collecting station for recycling newspaper, aluminum articles, and glass furnished one entire school with trays and glassware for science and art. The schoolchildren used the objects and then recycled them back whence they came. This is ecology in action. You and your students can practice what you preach.

The key to instituting a creative classroom is thinking about what better use can be made of particular objects or places in order to bring about an improved learning environment. This is not constant change for change's sake, but it is the application of thought to the solution of problems as they arise. This thinking involves the consideration of all the components of a room, school, or schoolyard. Try taking your class outdoors for science or to the gym for a mathematics lesson. Redesign your classroom using your own ideas.

THE LABORATORY
IN CREATIVE SCIENCING

The science laboratory need not be limited to an exclusive facility containing laboratory tables equipped with gas and water outlets, fume hoods, and associated laboratory hardware. The science laboratory may be located anywhere. The science laboratory is often a group of flattop tables in an elementary or junior high school. Modern schools do not require extensive facilities for their science investigations. Any facility will do — what occurs in that facility is MOST important!

One is not teaching science if laboratory experiences are not provided for the students. The laboratory provides the vehicle whereby students relate concepts and theories. Students can also use their observational skills to their fullest in laboratory situations. According to the National Science Teachers Association's *Theory into Action:*

The laboratory is a place to explore ideas, test theories, and raise questions. Here, meaning is given to observations and data. . . . The child's first experiences with science, beginning in the primary school, should involve aspects of experimental inquiry. He should learn how to observe with all his senses, how to measure, classify, use numbers, communicate, and practice similar subdisciplinary skills.... We need to recognize that the value of an experiment lies more in the means it presents for exploring the unknown than in the verification of the known.[2]

The role of the laboratory in science is extremely important. The laboratory should not be limited to filling in blanks in an activity booklet. Students should be involved in solving a problem or answering a question by collecting and interpreting data. Let's talk about some of the essential features of laboratory investigations.

A problem or question   We want, not low-level questions such as, What happens when magnesium (Mg) ribbon is added to hydrochloric acid?, but higher-level questions such as, How is the quantity of hydrogen gas produced related to the amount of ribbon that's present? The first question involves an answer that does not need a laboratory investigation, but the second does. Students will need to prepare a hydrochloric acid solution of known concentration and then react a constant volume of HCl with varying masses of Mg ribbon. Students will need to determine independent and dependent variables involved in the experiment. They will not know the answer to the experiment before they begin. They will collect data, use graphing techniques, and attempt to look for a pattern or trend in the data. Other systems such as Mg ribbon and sulfuric acid can be used to see if this trend can be generalized. The laboratory experience should be open-ended, not closed as presented in the first question.

A list of materials or supplies needed   The list should be held to a minimum. Only the simplest equipment that will do the job should be used. In many cases the equipment list should be left open so that the students can design their own — such as a collection tank for the study of the gas generated by an Alka-Seltzer tablet dropped in water.

Objectives for the investigation, preferably stated in performance terms
Objectives, such as a project in which the student can observe and calculate the rate of stem growth of plant X over a three-week period and then predict the stem length at the end of four weeks, provide enough student direction so that he or she can begin the experiment without

2. National Science Teachers Association, *Theory into Action* (Washington, D.C.: National Science Teachers Association, 1964), pp. 14–15.

waiting for instruction. The student may make mistakes in getting started or during the experiment, but this often happens when anyone undertakes any experiment. Students and teachers need to understand that

*there is not just one right answer
to an experiment or science activity*

Sometimes there are many acceptable answers, or answers that will vary because of experimental error in the instruments used, or human error. Students need to realize that their laboratory results are not perfect. Teachers should realize that the answers listed for experiments in the teacher's guide are not perfect either.

A list of references The student requires sources of background information for his investigation — textbooks and periodicals should be included on this list. Appropriate books from the school and local city or nearby college library might be included as well. Students should understand, and so should teachers, that they cannot rely on one textbook for all their background information and reading. They need to receive different inputs of information so that they can sift, sort, and reassemble the data collected in order to propose a solution for a problem.

Evaluation A generalizing experience should be included so that the students may apply the results of their investigation to other related laboratory experiences. The evaluations should be oriented to the activity for the student rather than being strictly paper-and-pencil evaluations. Many students will propose paths they would like to pursue for further study. These suggestions may serve as their evaluation of the investigation. For example, a student may suggest as an evaluation that our Mg–HCl investigation could be extended by using tin (Sn) or aluminum (Al) with HCl. Can the student predict the trend and analyze the data? Appropriate aids (audiovisual and others) should also be used with this experiment.

In a creative-sciencing laboratory, many of the routine techniques often demonstrated by the teacher can be handled through the use of film loops or programmed materials. Many film loops are available which demonstrate how to use a balance or how to measure volume. Many students have prepared their own film loops showing these techniques. Such loops are much more interesting and receive far greater use when the students' friends are involved in the demonstrations and production of the film. In this way each student can have his or her own demonstration showing how to use a piece of equipment when he or she needs to know how to use it.

In a creative-sciencing laboratory, fewer pieces of conventional science equipment are needed than in a traditional laboratory. As a result, the cost for equipping the creative-science laboratory is much less. Because of this, additional money is often available for other essential equipment that you would like to have. Not all students need to use a balance or graduate at the same time, when a variety of activities are available.

Creative-sciencing teachers realize that convenience costs money. Purchasing catalog equipment and commercial science-kits is tidy, convenient, and costly! By and large, you can realize tremendous savings by purchasing your own science equipment, materials, and supplies for making your own kits. You will also save money by purchasing components to construct equipment of your own design, such as circuit boards and buzz boxes. The flexibility in purchasing or acquiring materials through any means possible allows you to acquire exactly what you need in the quantities you need for your particular class. One innovative teacher in a large suburban school saved her district over $200,000 in one year by making equipment for districtwide use and having children help her assemble the materials.

Catalog shopping should be replaced by repeated tours of supermarkets, lumberyards, junk yards, garage sales, discount stores, drugstores, auctions, stationery stores, cellars, attics, school supply cabinets, and home closets. Department stores frequently can furnish you with discarded displays or advertisements that may contribute something to your instruction. Lumber companies are usually good sources of supply for remnants of wood, shingles, insulation, ceiling tiles, bricks, and other building materials. Drugstores occasionally are good sources of supplies ranging from cigar boxes to pillboxes to be used in insect collections. Industrial firms offer a variety of items that may be of value to you in your science program. Parents in the community are excellent sources of many needed items and skills. This list could easily be extended. The point is that many sources would be willing to contribute if you ask. Know what you want. Get nosey. Find out who has what you want, and then *ask for it*. Remember, the squeaking wheel gets the grease. Advertise, run an ad in the local paper, put a blurb in the next Parent-Teacher Association letter, start a telephone campaign. You will make a lot of friends and you will exceed your needs.

It is not always necessary to go beyond the confines of your school to get help. Try asking the principal. The administration stands in the most crucial position of all in regard to curriculum change or improvement within the school. Administrators usually are more than willing to assist you where and when necessary. Sometimes they need to be sold. This means a well-planned presentation by you of what you want to do, what it will cost, some financial alternatives, a comparison of commercially prepared materials versus your ideas, the implications for improving the science program, and what you are willing to put into the proposed innovation by way of time and effort. It is almost impossible for them to deny honest, enthusiastic initiative. Again, ask!

It does not pay to circumvent the establishment. Cooperation between teachers and the administration changes both parties for the better. People who set out to produce change must not only be tied up in the process themselves, but they must involve others. In order for a curriculum to change, people must change. Sometimes this is a slow, painful process but your acting as a creative, innovative model can do much to start the process moving.

Space requirements for sciencing can be a problem. But every problem has a solution. Again, the administration can assist you. But they must know what you want, how you plan to use it, and what significant results might accrue from their action. Often just storage space may be needed. Schools have space. It is not always used fully or wisely. Tour your room. Tour the building with an eye for renovating it in terms of needed storage space. In some cases, adding a storage building may be possible. Think big. You might have to sell hard, but if the end justifies the means, it will probably get done. No principal is anxious to see a good idea that will reflect well on the school, the teachers, and himself go down the drain. What is good for you will probably be good for him. There is space — you need to find it. Then *ask for it*.

Getting work space inside the classroom may simply mean rearranging things. Consider exchanging new desks for an old, low workbench that students can indeed work on. This work may range from drilling and sawing to soldering, gluing, and hammering. Active involvement in science requires a work area. This should be an area where at times a mess can be made, but which can later be restored to some semblance of order.

Classrooms should be more of a work area for students than an arena for spectators in learning. Students are often accused of acting without thinking. One possible explanation for this might be that we expect them to think without acting.

Many school supply closets contain science equipment for your laboratory that has yet to be discovered. Seek and you may find microscopes, beakers, test tubes, tuning forks — equipment of every variety — all yours for the seeking.

To succeed in creative sciencing inexpensively, adopt the substitution habit. Substitute a baby food jar or a peanut butter jar for a beaker. And:

When you need:	Substitute:
Aquariums	Gallon jars
Vials	Used pill bottles
Eyedroppers	Soda straws
Weights	Fishing sinkers
Spheres	Marbles
Dowels	Broomsticks
Trays	TV dinner trays, pie trays, oleo tubs, school lunch trays
Screening	Nylons
Scoops or shovels	Plastic bleach bottles

The best sources of supply that you have are your students. They can help you obtain the articles you need for your science laboratory.

A creative-sciencing laboratory allows a maximum amount of student involvement and a minimum amount of teacher direction. Students are free to select those experiments which are applicable to their goals as indicated by a prior evaluation or teacher guidance (not teacher dictation). The laboratory is intended to allow the student to be as creative and innovative as he or she can be in designing laboratory apparatus and conducting an experiment. With this opportunity for maximum involvement and creativity, the student will experience "real" science, not textbook or pseudoscience.

At the elementary school level, the basic skills of science should be emphasized in the laboratory using an array of content selected from all areas of science. The students should be concerned with acquiring the skills of observing and describing observations in both quantitative and qualitative terms. When observing, children should become aware of the need for involvement of all their senses and for extensions of their senses.

Beginning at the primary grade levels, children should be introduced to the metric system and become knowledgeable in its use. As stated earlier, they need to be exposed to a variety of science content from all areas, not just life science or physical science. Teachers and students should not try to place labels on all the science-content areas; they need to realize that they are working with science itself and that the fundamental aspects of science tie all these areas together.

Children at the intermediate, middle school, or junior high years should be introduced to the integrated skills of science. Their science experience should build upon the basic skills introduced during their primary years. These children should be involved with formulating hypotheses, making predictions from observed data, and formulating models on the basis of these data and their predictions. Again, experiments should be selected from all areas of science and the children should be aware that they are working in science, and not concerned whether that science is biology, chemistry, or physics! Science investigations at these levels should provide for a variety of activities. The students should be allowed the freedom to design their own equipment and select extensions of their experiments.

As you now realize, the approach you take to your science laboratory activities will serve as the key to whether or not your science program is successful. A creative-sciencing approach is built around a core of laboratory activities which emphasize student participation and involvement to the fullest. Students are involved with answering questions or solving problems to which they do not know the answers. These solutions and answers cannot be found simply by looking in their science textbook. Laboratory experiences emphasize process skills of science rather than content areas of science. Students are not competing with each other to see who got the right answer or who produced the most. They are, however, concentrating on finding the answers to many basic science questions and discovering or confirming many science laws by themselves. They are formulating models and using these models to see if new theories will evolve and if these new theories stand the test of time.

PAUSE FOR A SUMMARY

Successful creative-sciencing teachers *emphasize* alternatives to traditional science programs by means of the following:

— student involvement and participation;
— direct (hands-on) experiences through activities such as field trips and workshops;
— the use of homemade, simple equipment;
— the wedding of science to all curricular areas;
— an inquiry method of learning emphasizing process skills such as observing, measuring, classifying, and so forth;
— meeting the individual needs of each child;
— the concerns and feelings of the child *over* the content and one's own feelings;
— alternative classroom arrangements;
— teacher-developed materials;
— self-development through students' acquisition of their own materials;
— student-to-student as well as student-to-teacher interaction;
— open-ended questions which require a variety of responses;
— receptivity to new ideas and willingness to change;
— provision of experiences so that children can make independent decisions;
— use of a wide variety of materials.

CHANGE THROUGH INDIVIDUALIZED INSTRUCTION

Everyone is for it, few know what it connotes, and fewer know how to do it.

Individualized instruction is centered around the student — not the teacher, the school, the program, a computer, or a tape recorder. Individualized instruction should be for every student enrolled in your course, whether at the elementary school or college level.

individualized instruction is a specific
program of instruction for each student,
based on his previous experiences,
interests, and abilities

In individualized instruction, all the students do not proceed through the same materials at their own rates as they do with self-paced instruction. All the students do not proceed through the same taped sequence of materials as they do in audiotutorial instruction. Furthermore, all students do not proceed through a sequence of frames with immediate answer confirmation as they do in programmed instruction or computer-assisted instruction.

Individualized instruction involves student-teacher planning (the order is important) of materials based upon the entering behaviors of each student. Individualized instruction involves teachers' preparing individualized materials for individual students.

An excellent way to begin individualizing your creative sciencing is to write down what you plan to offer your students as a program of instruction. Use action verbs which emphasize skills such as observing, classifying, predicting, measuring, and so on. Make sure that you don't limit your list simply to the areas of knowledge and recall. Your goals or objectives should also include those drawn from the affective and psychomotor domains; Section 6 may provide some assistance.

With your set of goals and objectives, you must ask yourself if all your students will need to proceed to all the goals and objectives you have delineated for your creative-sciencing endeavor. Do some of your students already have many of the capabilities listed? If, for example, one of your objectives is to be able to measure the density of an object to the nearest tenth of a gram per milliliter and four students can do this, should they have to do it along with the rest of the class? Could their time be better spent working on some of the other objectives you have prepared for your course? We would think that it could. As a result, you can see the need for an evaluation of your selected goals and objectives. This prior assessment will enable you to identify, before instruction begins, those students who can adequately perform specific portions of your creative-sciencing program.

Whether or not you realize it, through the process of constructing goals, objectives, and preassessment evaluations, you are already beginning to individualize your instruction. The next step is to meet with your students (they should have a part in deciding what they need to do and how they must proceed in order to acquire creative-sciencing skills). Together you should prepare an instructional outline or sequence of activities or procedures that they will need to follow in order to master the material presented. But that's still not individualized instruction. If you think it is, reread the definition. Does this alter your decision? The interests and abilities of each student still need to be taken into account. From the preassessment measure, you have an idea of an individual's background experiences in science. Meeting with each student for a general discussion as well as examining the student's past record in other science and related classes will provide additional information. An attitude inventory as well as past performance on standardized tests (whether the Sequential Test of Educational Progress at the elementary level or the Iowa Tests of Basic Skills at the junior high level) should provide you with a good cross section of the student's background experience, interests, and abilities. As a result, you and the student should be able to begin to make some decisions about performance in your creative-sciencing classroom. The student may have already met some of your goals and objectives, and others may not be applicable. The list you make now is this individual's, and probably no other student in the class will have the same list. Now you're on the road to individualizing instruction. Are you getting uneasy? Will this take too much time? Remember:

nothing is as hard
as just getting started!
keep reading!

Many of your students may be experiencing their first contact with an individually prepared creative-sciencing program. Consequently, they may be confused at first. Be patient. They will miss meeting every day as a group with a teacher who tells them what to do each step of the way in a lockstep learning environment. They may feel uncomfortable in having individual programs of study. The weight of the transferred responsibility can cause difficulties. An individualized course at a higher grade level can cause confusion and frustration among the students. To assist students in the changeover to individualized instruction, many teachers have used student contracts like the one on page 58. They are truly contracts, signed by the student and the teacher and dated. These contracts specify when certain parts of the individualized program will be completed and turned in for either teacher evaluation or self-evaluation or both. Students state that these contracts help them keep track of their direction and serve as a valuable substitute for daily teacher assignments. Most students who have progressed through one individualized course will probably not need contracts in future individualized courses of instruction.

You will find it to be helpful if you ask students who have completed your individualized creative-sciencing program to return and assist you as student guides. Laboratory assistants can serve the same purpose as student guides. They can assist with the teaching that is needed (often doing an excellent job) as well as setting up materials and equipment. Furthermore, student guides serving as part of the learning team can provide an effective learning environment.

If you're still with us, but not yet convinced about all this — don't despair! For individualized instruction to be successful in science teaching as well as in all areas, the roles of the teacher and the student in our classrooms must be modified. Teachers are no longer the science experts or lecturers; instead, they are guides. They guide individuals. They are the stage setters for learning. Teachers who teach an individualized class often remark that they now know their students better than they have ever known students before. They know their background experiences, interests, abilities, moods, whims, and life goals. They also find that the students have a greater interest in science — it's no longer boring and remote from their daily lives. However, teachers who teach individualized classes need time — time to meet with students individually, time to prepare materials (many different types), time to evaluate their students' progress, time to work with student guides, and time to think and be creative.

Good teachers of individualized science are more important than facilities. Most existing facilities are now adaptable to individualized science procedures. A variety of areas in the room is needed as well as a good instructional-materials center in the school, where students can work independently.

What about the student's role in an individualized class? The student is no longer a passive learner who sits back in class and daydreams for one reason or another. Maybe the student who daydreams either knows the material already or can't understand the material because it is not being presented at his or her level. Students in an individualized class are using materials designed for them and that they helped to design. They become active learners. They are using materials they can cope with. The superior students in science can go as far and as broad as they wish. The less able students can work on those areas of science in which they are most in need of assistance as well as pursue their own interests. With the trend toward mainstreaming, special-education children can easily participate in the creative-sciencing class too. This also takes time.

A majority of students will not *cover* as much material as you may have covered in the past. These students, however, will have a greater

STUDENT LEARNING
CONTRACT

I _____, agree to successfully complete the
<small>(STUDENT'S NAME)</small>
following activities and objectives on or before _____.
<small>(DATE)</small>

Objective or activity

1.

2.

3.

4.

5.

Signed _____
<small>(STUDENT)</small>

Approved _____
<small>(TEACHER)</small>

Today's Date _____

knowledge and understanding of the material they do cover than they
would have had if taught by traditional methods. If one of your goals
is to cover lots of material, individualized instruction may not meet that
goal. Students state that they have a greater responsibility for learning

in an individualized science course than in the traditional course; that is, the main responsibility for learning is now placed upon the student's shoulders and not on the teacher's.

Critics of individualized instruction state that many students waste classroom time — "They don't *do* anything" (according to these observers)! There are two reasons for this impression: first, students new to an individualized science course often do not know how to handle this new freedom for learning, and as a result they need assistance (student contracts help to get them started); second, students probably aren't wasting any more time than in any other class, but it just seems more obvious to an observer of an individualized class.

An efficient way to assist teachers and students to adapt to their new roles is through the development of modules, interest centers, or packages of materials developed by teachers. Students can use these as part of their science program (as discussed in Section 3). There are many materials on the market at the present time that can be adapted for use in modules and interest centers. Many of the national curricular programs in science (see Appendix B) have components that can be used for interest center and module development. Don't worry if you can't think of a module or interest center on your own — most people initially have difficulty preparing these. You need to collect many representative materials that are available in science for your grade level or science subject area. From these you can select those materials that may best fit your needs and then adapt them for use as your modules or interest centers. Most of your students will be able to assist you with the development. They might suggest areas they wish to include as part of their science study or areas where individual packages may be of assistance to them.

Teachers who attempt alternative creative-sciencing approaches often have the problem of trying to keep track of the progress of each child and the materials with which each is involved.

KEYSORT KID CARDS

The Keysort Kid Card method will enable you to keep track of each child — to know immediately who needs additional help or who has successfully completed an activity or who hasn't. Keysort Kid Cards take up little space and little teacher time.

To use Keysort Kid Cards, you will need to prepare a set of cards. One is shown on the next page. Copy or duplicate the card and tape or glue it to a piece of cardboard or manila folder cut to the same size (computer keypunch cards also work well). You will need at least one copy for each student in your class, so be sure to make lots of duplicates.

Forty punch-holes on this card will allow you to use up to forty topics or evaluation ideas for each child (if you use two cards you will have eighty). Write the name of each child in your class on a separate card (thirty children, thirty cards). When a child completes an activity, the hole for the corresponding activity number on the card will be punched out. Be careful *not* to punch through the edge of the card (or through the numbers).

Let's consider a typical third-grade class of twenty-eight children. Keep in mind that virtually any subject or topic can be used. You are limited only by your creativity and your imagination.

Suppose that you have a list of twenty-five science activities and that you want each child to successfully complete at least ten of them. These twenty-five activities can be represented by the first twenty-five numbers on the card. When Jerry, for example, successfully completes his first activity, punch out the hole *through* the margin of his card for number 1 as shown in the illustration.

A knitting needle or any similar object that will fit through all the holes of any stack of cards will assist you in determining who has completed what activity and how many activities have been completed by each child. The cards of the students who have not completed a particular activity will stay in the stack. Another excellent feature is that if a small group discussion relating to activity 4 is desired, the teacher can pass the knitting needle through the fourth hole of all the class cards. Those cards that drop out have successfully completed the fourth activity and are ready for the discussion.

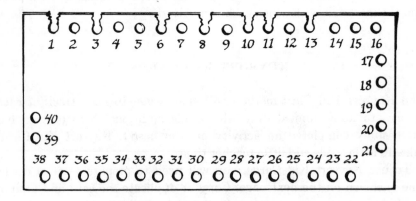

Teachers also use this method for keeping track of children with reference to reading and mathematics skills, group work, and for evaluation purposes. A commercial version called "Tracking Cards" is designed to go with *Science — A Process Approach* (which is listed in Appendix B). You will probably have many additional uses for this idea.

CREATIVE-SCIENCING AIDS

The volume of aids available for creative-sciencing alternatives is overwhelming. The mere use of an aid, for example, a cassette tape recorder, does not constitute an alternative science program, however. This section will give you a representative list of common aids that can assist you and your students in your creative-sciencing endeavors. Many of these aids can serve as substitutes for the real thing when it is impossible or impractical to bring the real thing into your classroom. Animals such as the white whale, elephant, seal, octopus, squid, or lion are fascinating subjects for children; most principals would cringe, however, should you requisition hay to feed your class pet, the elephant!

There are excellent media aids available that will do the job. Super 8 film loops are outstanding for individual or small-group use. Loops which duplicate experiments that would be difficult for you to duplicate and loops that feature microscopic animals, like brine shrimp, are recommended. If you can easily duplicate what the film loop does, then, of course, you needn't buy or rent that loop.

Many teachers and students enjoy preparing their own loops. This technique provides the student with direct experience in photography preparation and use of film loops. They are less expensive than commercial loops, and student-prepared loops may be more closely related to the materials being used in your classroom. Most photography stores would be happy to advise you and your children on how to prepare a film loop.

Computer-assisted instruction (CAI) is rapidly making inroads on our school programs. Many schools are renting access terminals from nearby colleges and universities in order to provide CAI experiences for their students. Schools are also purchasing computer-calculators to replace the long and tedious workbooks for mathematics. We have not yet begun to comprehend the potential that exists in this area. CAI can be utilized with each individual student and can serve as a teacher's aid by taking care of many bookkeeping duties such as attendance, individual student progress, and evaluation measures.

Computer-calculators can assist students in solving many problems in science and mathematics that often take long hours of tedious, mechanical manipulation if done by hand. Students should receive instruction of a hands-on nature in preparing computer programs and in using computers. Remember, their world is not our world and they will be living in a world vastly different from today's.

Audiotapes, either reel-to-reel or cassette, have been used successfully in assisting students in science as well as all other curricular areas. Almost any school-age child can successfully use a tape recorder after a short introductory session. Audiotapes allow you to provide for the individual needs of your children. Audiotapes have been found to be useful in assisting children with reading problems. The narrative is recorded and children can read along with it. Each child can have a science story (it's the number one topic that children of all ages prefer) presented on his or her own reading level. The child need not be penalized for not understanding the science story because of its original difficult reading level. On the other hand, students at a higher reading level can be presented with science stories in their original form. After a few years, you will have developed an excellent library of science stories at a variety of reading levels for use with your students. By the way, children enjoy preparing story tapes for other children — it also helps them improve their reading and speech skills.

Videotapes, especially with portable video cameras and recorders, are excellent media aids. Portions of commercial or educational TV science programs can be taped for later use by your class. Students can write, produce, and record their own science shows using portable video cameras and recorders. Teacher demonstrations can be videotaped, allowing you freedom to work with individual members of the entire class. The future will find a much greater use of the videotape capability than we are presently achieving in our schools.

Sound films (16mm) are also excellent media aids. However, they are probably the most abused aids in use today. Unfortunately, most of the so-called science films are more nearly encyclopedic presentations stating either all the parts of a grasshopper in fifteen minutes or showing unimaginative, complex science experiments conducted by a man in a white coat. Only those films that *add* to your creative-sciencing endeavors should be selected. These films should include one or more of the following features:

— They should present the children with a problem to be solved.

— They should have children in the film, when appropriate.
— They should use simple, everyday, homemade equipment when possible.
— They should cause you to think by asking questions, some of which are not necessarily fully answerable.
— They should encourage discussion by provoking children to want to discuss the film at its conclusion.
— Finally, the films should present science as it really is and the scientist as a real person, not as the man in the long white coat.

Finally, let's consider the aid that every teacher uses (and abuses) — the *bulletin board*. Many school bulletin boards are merely that — boards for bulletins, made by teachers mainly to please other adults. We like to refer to bulletin boards designed for children as "action" bulletin boards. Action bulletin boards have the following characteristics:

— They often present the children with a problem in the form of a question.
— They involve the children by use of task cards or by having the children contribute to the bulletin board.
— They are continually changed as children perceive that they need changing.
— They are often tied to an interest center.

Are your bulletin boards action bulletin boards? We've provided an example of a real action bulletin board on page 64. Compare it to your own!

PAUSE FOR A SUMMARY

— Creative-sciencing teachers are always open to change, particularly when the change may improve learning for children.
— Creative-sciencing teachers need to expand their role as educational agents for change and continued improvement of their school, philosophy, attitude, objectives, and curriculum.
— Successful teachers are good questioners. They promote open-ended questions rather than closed questions with single-word answers.
— Creative-sciencing teachers encourage student-to-student interaction and allow sufficient time for student responses to questions.

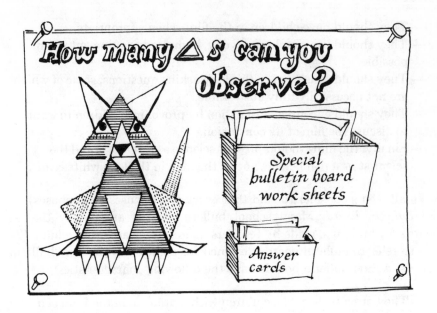

How many △s can you observe?

Special bulletin board work sheets

Answer cards

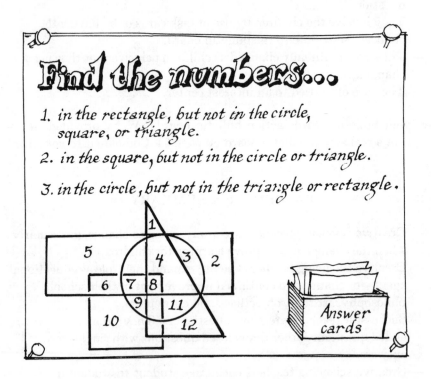

Find the numbers...

1. in the rectangle, but not in the circle, square, or triangle.

2. in the square, but not in the circle or triangle.

3. in the circle, but not in the triangle or rectangle.

Answer cards

— The key to a creative classroom is thinking about what better use can be made of particular objects or places in order to improve the learning environment.
— Laboratory experiences are the vehicle whereby students apply concepts and theories to reality.
— A creative-sciencing laboratory allows for a maximum of student involvement and a minimum of teacher direction.
— Individualized instruction is a specific program of instruction for each student, based on his previous experiences, interests, and abilities.
— The Keysort Kid Card method enables you to quickly assess each child and to know his orientation in discrete areas of the curriculum.
— Creative-sciencing aids include: Super 8 film loops, computer-assisted instruction, computer-calculators, audiotapes, videotapes, 16mm films, and bulletin boards.

Alternative skills
and techniques

Teaching children to teach themselves

Give me a fish and I eat for a day. Teach me to fish and I eat for a lifetime.
— Traditional

Karen: Science is pervasive. It reaches further than any other discipline into humanity's highest needs — to know and to understand. I have sat through far too many "science" classes in which my own need to know and to understand was smothered by statements like the following: "Turn to page 84 and fill in the blanks"; "Now here is how it works, class"; and "Save that question until I am finished explaining this." My philosophy of teaching science has thus evolved into one of simply trying to fulfill a child's need to know and to understand. My goals are to allow each child to explore, to manipulate, to goof sometimes, to get confused, and above all, to be a problem solver and not a teacher pleaser. There are many methods by which a teacher can achieve these goals. The lessons can be individualized in the form of learning modules or learning centers which provide plenty of activities and materials to manipulate and explore. The children must be allowed to handle the equipment, to do their own experiments, and to pursue their own goals. Teaching should not revolve around a textbook, but should be concept oriented. The teacher must learn to be a resource person and an adviser where needed. I have an image of a good science teacher as being like a wise pack rat — always on the lookout for some useful junk and always wise enough to know how to help children follow their curiosity wherever it leads them. All of the above methods are good ways to help a child fulfill the need to know and to understand. As usual, the catch is in the doing. It's easy to say glorious things about how to teach science, but it's another matter to actually discipline yourself as a teacher to avoid the pitfalls of authoritarian teaching methods and step up to a higher level of learning with the children.

Al: I believe that it is important to teach science in the elementary classroom for various reasons. Many skills developed through the teaching of science are extremely important in other areas of school life and learning. We might say that science, when properly taught, can help students develop skills in observation, discrimination, description, exploration, invention, discovery, and ability to discuss ideas and concepts.

These skills must be approached sequentially. Children must begin with the concrete and advance to the abstract as indicated by Piaget's research.

When science is taught with care and understanding, I feel that it can help to develop more proficiency in the areas of reading, writing, language development, development of the senses, and math.

Science, more than any other academic subject, can help students become more aware of themselves and their environment. Proper teaching of science can bring about an excitement for life better than any other subject taught in the elementary school.

As Karen has said, "The catch is in the doing." It is deceptively easy to command: "Become involved with your students, have lots and lots of activities for them, use their ideas, develop interest centers for them, individualize instruction, . . ."

but

Where do you obtain all these ideas?

Where can you find examples?

How can teachers develop alternative skills and techniques?

This section of *Creative Sciencing: A Practical Approach* is designed to offer you ideas and approaches for use in your classroom. These ideas and techniques have been used by children at various levels from kinder-

Alternative Skills and Techniques / 69

garten through the ninth grade. As starters several examples will be included. Once you get the idea, you will undoubtedly generate many ideas of your own.

Many of these short skills and techniques can be utilized as part of an interest center or instructional module. You are encouraged to duplicate these activities and approaches. Try them with children in the classroom as practical alternatives or supplements to the existing program.

TASK CARDS

An excellent way to begin involving your children individually or in groups in a variety of creative-sciencing endeavors is through the use of task cards (sometimes they are called idea cards, suggestion cards, or activity cards). Many teachers like to select a theme, such as pollution, and then prepare from fifteen to twenty task cards dealing with this topic. Task cards usually follow the basic form described below.

1. The front side
    a. The task card's title is in clear view and presents the topic of investigation. For example:
       — Air Pollution
2. The reverse side
    a. A statement, sometimes in the form of a question, indicates the task to be solved. For example:
       — Collect evidence from your community relating to the smoky polluters.
    b. You may give some suggestions on how to complete the assignment. For example:
       — Compare factory pollution to school pollution.
       — Design your own smoke chart.
       — Use a camera to record your polluters.
       — Make drawings of smoky polluters.
       — Prepare a letter to the president of a polluting company to find out what steps his company is taking to prevent current or future pollution problems.
       — Collect evidence by using Vaseline cardboard discs located near the pollution source.

CAN NOISE POLLUTION
BE FOUND IN YOUR SCHOOL?

TASKS

Tape record noises in and around your school.

Play your tape to other children. Can they identify the noise source?

Can you suggest ways to reduce noise pollution in your school? Home? Community?

Take (or draw) pictures of noise pollution problem areas.

MORE

Locate noise pollution problem areas in your community. Design a procedure for correcting them.

Measure the noise pollution with a tape recorder volume indicator or decibel meter.

Can loud noises hurt your hearing? Interview an ear doctor and then a jackhammer operator or jet engine mechanic.

What would you do if your school were located on the approach to a busy city airport? Design and implement a program of action.

Investigate the laws that have been instituted to reduce noise pollution. Try the Noise Control Act of 1972. Are there local city laws on noise pollution?

Did you know that in New York City honking your auto horn except in cases of "imminent danger" costs $50.00 a toot? Is this a good law?

c. A collection of additional, related activities titled "More" can also be included. For example:
   — Are smoky polluters polluting our environment in other ways? See if you can find out.
   — Prepare a pollution poem and present it to the class.

Many teachers prepare a task-card area in their room for student use. The task cards are changed periodically. Students are encouraged to write some of their own task cards. The next several pages include sample task-card ideas. Please look at them and use them with your students. Task cards are not bounded by age, grade level, interest, and so on. They work well with all ages of children (and adults too).

MORE THEMES FOR TASK CARDS

1. Go through a magazine and tear out all the pictures of things whose names begin with the letter C. Paste them all together on a piece of construction paper.
2. Collect animal pictures. Then choose an animal that you are interested in and find out more about it.
3. While you are at recess, collect some things that interest you. When you come in from recess, paste them on a piece of tagboard.
4. Find a picture of something you enjoy. Make up a commercial for it. Now try to sell this product (or thing) to your friends.
5. Count something that you have always wondered about. For example, how many chairs and tables do we have in our room?
6. Take a picture of what you dislike at school and decide how you might change it. When you have decided, tape-record your suggestions.
7. Go through a magazine and tear out five pictures that you like. Make up a story about these pictures. When you are sure of your story, go to the tape recorder and record it.
8. Trace this shape by walking:

   Trace this shape by jumping:

Trace this shape by hopping:

9. Using old magazines, cut out five long *o* words and five short *o* words and paste them on paper as a decoration.

10. Find a recipe for butter, yogurt, buttermilk, ice cream, or any other milk product which is usually sold in stores. Then make it. When it's done, taste it and decide what you did right and what (if anything) you did wrong.

11. Set up an experiment to find out what conditions affect carbon dioxide (CO_2) production in a mixture of sugar, yeast, and water. Do not use more than four pounds of sugar or one-fourth of a package of yeast for any one trial.

12. How much allowance do you get each week? Do you have a job? Write down how much money you get each week and estimate how much you spend and what you spend it on. Then keep a record for a week of everything you've earned and spent. Does the record agree with what you estimated?

    MORE: Now *plan* what you'll earn and what you'll spend for another week and try to follow that plan. Such a plan is called a budget. What can you budget besides money?

13. Borrow a basic cookie recipe from your mother and then bake the cookies. Did you follow the directions exactly? What would the cookies look like if the oven were hotter? Or cooler? Would you have to adjust the time? What would happen if you put in too much baking powder? Or too little flour?

    MORE: What spices would make the cookies more exciting? Cinnamon? Nutmeg? Ginger? Would you add brown sugar? Coconut? Nuts? (How much?) Experiment and create your own cookie recipe.

14. Look up some linear equations in your algebra book. Graph the equations. Make a Christmas tree by putting dots on

paper and then connecting them. Could you write equations for all the lines connecting the dots? Try it.

MORE: Try this on your classmates. Give them a set of equations. The one you figured out for the Christmas tree will do. See if they can draw the picture. Try writing equations for other pictures, such as a star, house, or snowman. (Can you use only linear equations for a snowman?)

15. Using magazines and newspapers, make a display of a variety of geometric figures as they occur in architecture, nature, science, advertising, and other fields.

16. Pretend that you are an immigrant in a foreign country. List the problems you will probably encounter.

17. SCIENCE MISSION POSSIBLE (SMP) NO. 1
Your mission, should you decide to accept, is on the card before you.

It will be your job to locate the following items placed in inconspicuous areas of the room and accurately measure their length in both the Imperial and metric systems. If for any reason you are unable to do so, you may contact an associate, but not under any other circumstances. When you have completed this assignment, meet with your superior before going on. Take precautions, because within fifteen seconds this card will not self-destruct.

The items are:
— your teacher's umbrella
— a Kleenex box
— a filing cabinet
— a bobby pin
— the chalk tray
— a 1976 magazine
— three science books in a straight line
— a one dollar bill
— a paper clip
— the demonstration table

If you feel you have successfully completed this assignment, then state in writing what you think is the general difference between length and width. (Materials needed: meter stick and yardstick.)

18. SMP NO. 2

The heat is on now . . . the word is out. It is our job to work as a team and cool it. Check your thermometers. What do they read? Not sure? Then do this.

Using your special SMP Celsius thermometer, find the temperature of the following:

— the air in the room
— the water from the faucet
— boiling water
— ice water
— 10 ml of alcohol

Now find the average temperature of the faucet water, boiling water, and ice water. Can you operationally define temperature? If so, we are ready to take on the job. Materials needed:

— 10 ml graduated cylinder
— water
— safety glasses
— 100 ml beaker
— ice
— alcohol
— alcohol burner

19. SMP NO. 3

The United States Government has very rigid restrictions on gold standards. There is reason to suspect that there has been some foul play. It is your job to locate the problem and expose the people involved. It is necessary for you to have a good understanding of what is meant by a standard of measurement. In order to prove your understanding, set up your own system for measuring length and weight. In other words, make your own units, but they must fulfill the requirements for a good standard of measurement. (Materials needed: They are up to you.)

20. Open a can of alphabet soup and spell as many words as you can with the letters you find in it. Write your words on a piece of paper.

21. On a piece of paper, write as many words as you can find in the word *astronomy*.

22. Look at a map of the United States and draw or trace the outline of one state. Now, turn your paper, looking at the shape until you "see" another object in it. Draw the new object.

23. Invent three animals by combining body parts of different animals that are familiar to you. Name them and draw a picture of each one. For example:

    elephant + hippopotamus = hippophant

24. For twenty different letters of the alphabet, find an object in your house or outside of it whose name begins with each letter and make a list of them.

25. Animals cannot talk, yet we can tell how they feel by the way they act. How do these animals act:
    — a horse, when it is scared;
    — a dog, when it is hungry;
    — a kitten, when it is hurt;
    — a bear, when it is mad?

26. You are in charge of pizza for our next party, and we want something exciting! Create a new variety of pizza and try it out; then we will make it at school. (Hint: Must all of the ingredients bake the same amount of time?) What would you put on a Christmas pizza? A Mexican pizza? Is an ice cream pizza possible?

27. Design your own bug catcher. Show the class how it works by setting it up outside the classroom. Use any materials

you like as long as it catches live bugs. (Dead bugs are no fun to watch.)

   MORE: Can you design a trap that catches only small bugs? Only large bugs? How about catching only flying bugs? (Hint: How can you attract the bugs?)

28. Find color photos of a variety of specimens of butterflies and moths. Try especially hard to find Monarch, Viceroy, Kallima, and *caligo eurilochus*. Do some of the wing colors and patterns seem useful? Find some examples of camouflage in other animals. (Hint: Imagine a polar bear hunting a snowshoe rabbit in a blizzard!)

   MORE: If you could design your own butterfly, how would you design the wings for protection? Where would it live?

29. Collect at least ten different kinds of flowers (or pictures of flowers where you can see all of the petals). If the flower has just a few petals, count them. If it has a lot of petals, count the sepals. (What's a sepal?)

30. Write a letter to someone you wish you knew or you admire, like a movie star, athlete, author, political figure, or scientist.

PUZZLER ACTIVITIES (PA)

Let's pause for a puzzler activity. Throughout the rest of this section of *Creative Sciencing: A Practical Approach* we will pause from time to time to present you with a PA (that's short for *puzzler activity*).

PAs are designed to help children learn to think by presenting them with problems to solve. PAs are brief confrontations with real or fictitious situations supplied by the teacher in the form of dittoed handouts. Challenged to solve the problems, the students may map out strategies, raise appropriate questions, and consider differing explanations.

Each student can contribute to this activity because no prerequisite knowledge is usually needed. Involvement in these activities supports the creative venture that is science. The PAs do not have any single right answers; some have no answers at all. PAs with no apparent answers can still be used — you can ask, "What questions, if you had the answers, would help define the explanation?" You may furnish additional facts to maneuver the discussion to a conclusion or by asking, "What more do we need to know?"

The student's response should always be accompanied by a suitable reason for the response. You as the moderator, or better yet, a child as the moderator, can challenge or accept each response as you promote thinking through the problem.

PAs can be a valuable adjunct to your sciencing in the classroom. They can also aid your other programs, such as reading. Try writing your own PAs by looking for source material around the house, the school, or in newspapers. Look for interesting and puzzling activities!

PA 1
WHAT'S ALL THE NOISE ABOUT?

Virtually all of the children given hearing tests at the Dekro Middle School in Indiana, New Jersey, are found to have some hearing loss. Many of the children are also emotionally upset. It is not uncommon for the children to get into fights with one another. The teachers are unhappy with their working conditions and are very vocal about it. Similar events in two other schools in the area have led to their closing.

Some questions to think about:
  a. Is this a school for children with special problems?
  b. Are these problems the results of something that happens at the schools?
  c. Is there tension between the faculty and the students? Between the faculty and the administration?

Write down what you think the problem might be here:

_____

_____

_____

_____

_____

_____

_____

What additional information would you like to have?

_____

_____

_____

_____

_____

Now consider these additional facts:
  a. **Dekro Middle School is located near the Indiana International Airport.**
  b. **Decibel readings range from ninety-five to one hundred in the school yard and eighty to ninety-six in the classrooms.**

Does any of this new information change your first conclusions? If so, what do you now think the problem is? Can you suggest possible solutions to the problem?

THE MINIPREPARATION
OF AUDIOTUTORIAL TAPES (MINI-PATT)

A procedure that allows teachers to produce instructional tapes suitable for their own classes in a short period of time (one to three hours for a thirty-minute tape) is now available using the Mini-PATT approach.

To prepare instructional tapes using the Mini-PATT system, just follow the directions.

1. Materials needed:
   a. One cassette tape recorder with microphone
   b. C-60 cassette tapes
   c. A partner, preferably a teacher or child
2. Procedure:
   a. Decide what concept you would like to teach using the audiotutorial method. For example:
      — the letter *o*, graphing, short vowel sounds, shapes, colors.
   b. State your instructional objectives for the concept you've selected. For example:
      — The child will be able to construct a two-dimensional graph from the data presented using acceptable graphing techniques.
      — The child will be able to observe an object using four of his five senses.
   c. Obtain all the materials and supplies that you will need to implement your instructional objectives. For example:
      — an object for observing or a game you've constructed
   d. Prepare an evaluation for this concept, based on your stated objectives, to be used later to find out if your audiotutorial lesson was successful. Do *not* show this evaluation to your partner!
   e. Using the materials needed, teach your partner the concept and record your instructions on tape. Your partner should respond by gestures or movements of the head, *not verbally*. Your lesson should last about fifteen to twenty minutes.

Adapted from Jimmy R. Jenkins and Gerald H. Krockover, "The Mini-PATT Approach to Individualizing Instruction," *Science Activities,* vol. 10, no. 3 (November 1973), pp. 38–39. By permission.

PA 2
WHERE WOULD YOU LIVE?

You have been asked to build an apartment house for thirty-two families in the town of Smokeville. About fifty thousand people live in this town. Smokeville has three large factories. The apartment house you are to build is one-quarter mile away from the steel mill. Here are some facts:

— About seventy out of every hundred people in the United States live in urban and suburban areas.
— City streets have become very noisy.
— Office buildings have become very big and noisy.
— In small towns, sound levels during the day are about eighty decibels.
— The sound level during the day in New York City is about ninety-two decibels.
— Sounds above ninety decibels can cause a person to lose hearing over time.
— Some common sound levels are:

Refrigerator	30–40 db
Dishwasher	60–80 db
Air conditioner	70 db
Vacuum cleaner	70–85 db
Disposal	90–100 db
Heavy street traffic	85 db
Factory equipment	85–130 db

Which of these facts would be of importance to you? What other information might you need before building your apartment?

You want to keep the noise level in your building very low. How would you build your apartment? What materials would you use? *Think!* What kinds of appliances would you put in each apartment? Would you air-condition the building? If so, would you put an air conditioner in each apartment or would you have one large one for the whole building? What other things might people have in their apartments that would be noisy?

Suppose you could build your apartment any place in town. Which spot would you choose? Why? How would you change your building plans if your apartment were going to be put in that part of town?

f. After completing your audiotutorial tape, listen to the results. Record your comments regarding your product for future revision. Both positive and negative aspects should be included.

The Mini-PATT approach is an effective procedure for developing audiotutorial tapes for classroom use. Many of the Mini-PATT tapes can be incorporated into interest centers, learning activity packages, and minicourses. Will you try the Mini-PATT system in your classroom?

INVITATIONS TO INVESTIGATE

Presenting children with a problem for investigation can be difficult. It is difficult because teachers are confronted with children who have a wide variety of skills and abilities even though they are assigned to the same grade level. One way to overcome this problem is through the use of an "invitation to investigate." An invitation to investigate may be defined as a problem-solving approach wherein the students analyze and synthesize.

Let's look at several illustrations of invitations to investigate.

Where did the water go?

Using two test tubes, rubber stoppers, and glass tubes as shown on the next page, prepare the following demonstration.

Fill the tubes so that they have identical warm water levels. Add rock salt to tube B and push the rubber stopper in or out to adjust the water level until the water level of tube B is identical to that of tube A. Ask the children to predict what will happen. Usually three predictions will be offered:

1. The water level will *rise* as a result of the melting of the rock salt.
2. The water level will *lower* as a result of the melting of the rock salt.
3. The water level will *not change* because the rock salt will not melt.

Based on our experience, most children will choose the first prediction.

Adapted from Gerald H. Krockover, "Invitation to Investigate," *Science Activities*, vol. 8, no. 5 (January 1973), pp. 38–39. By permission.

But, after approximately one hour, the children will observe that the water level will have gone down!

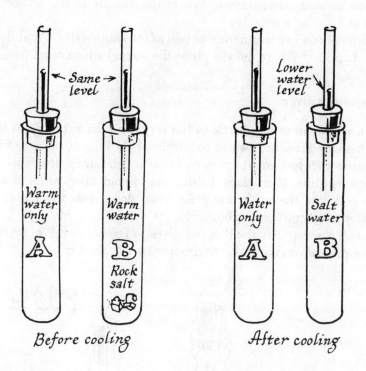

Before cooling   After cooling

Where did the water go? Children may now formulate hypotheses as to what has happened. Did the water level go down because a gas was produced? Or did the water evaporate? Did the salt absorb the water? Did the water absorb the salt? See if your children can design experiments to find out where the water went.

Further investigations might include trying table salt or sugar instead of rock salt. Are the same results obtained?

Are the volumes of any two liquids additive?

Try equal amounts of methanol and water (use a 100 ml graduated cylinder). You may need to experiment with a variety of quantities.

Try other liquids such as rubbing alcohol (isopropyl alcohol) and water or motor oil and mineral oil.

Cool it!

Inflate two balloons to the same size. Place one in the refrigerator and leave one at room temperature. Where did the air in the refrigerated balloon go?

Place a balloon in a warm oven or pan of steaming water. What do you observe happens? Where did the air in the warm balloon come from?

The Cartesian diver

Fill a clear wine or catsup bottle to the very top with water. Set it aside. Fill a drinking glass with water to within ¼–½ inch of the top. Fill an eyedropper with just enough water so that it will barely float upright in the drinking glass. Then place the dropper in your filled bottle of water. Place a cork into the bottle and press down slowly (see the figure). Excess water may squirt out, so be careful.

You will observe that the dropper dives to the bottom. When you release the pressure on the cork, the dropper rises to the top.

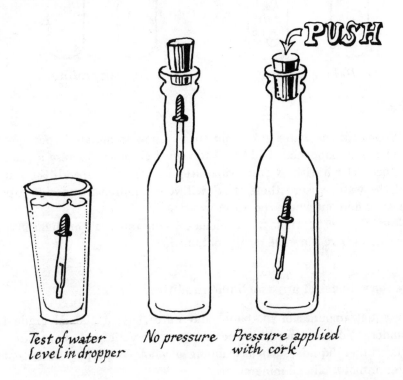

Test of water level in dropper   No pressure   Pressure applied with cork

This activity can serve as an excellent invitation to investigate. What causes the eyedropper to dive? Air pressure? Water pressure? Why does the dropper return to the top of the bottle when the pressure is released?

An extension of this invitation to investigate involves the use of a small, wooden safety match. Cut off a small segment (7 or 8 mm). This segment should include the head and a small portion of the wood behind the head. You may need to experiment a little to determine the proper length required. Place this small match bit into the water bottle. Press firmly on the cork. The match segment reacts to the pressure in the same way as the dropper.

Is the explanation for the behavior of the match and the dropper the same? Are the same components involved?

PA 3
WILL YOU JOIN THE TEAM?

You have just been appointed chief investigator for TEAM — Tippecanoe Environmental Action Management. Your job is to track down the culprits involved in the pollution and misuse of our environment.

You have been handed a report entitled "Wabash Crisis." As you examine the report, the following important facts seem to stand out:

— Water samples were taken along a one-mile stretch of the Wabash River on either side of the crossing of US 52 and Indiana 43.

— The samples were taken every other day for a period of two weeks, with no noticeable change in the content of the samples.

— Large masses of dead fish were found at the same location.

— The laboratory report on the samples found that the phosphate levels were above the danger point for a good oxygen supply; the mercury content of the fish was high; oil content was at 1.5 percent; and there were observable traces of organic wastes.

What steps are you going to take? Is there immediate danger to humans? How soon will the effects be felt? How many people and other living organisms will be affected? Can the situation be improved? Outline your procedures and your tactics.

SKILLETTES

Skillettes are short, discrete minilessons designed to introduce or reinforce a single skill. We can use the process skills of science (observing, measuring, inferring, predicting) as the basis for a series of skill lessons.

For our example, let us select the skill of classifying and illustrate the skillette concept with a series of classifying skillettes designed for the primary grades. We have prepared six skillettes to reinforce the skill of classifying. Each skillette includes an instructional objective and several activities to go with it. All the children may need to do all the skillettes; however, an initial preassessment would allow you to determine which children should use what skillettes.

To reinforce beginning reading in the primary grades, you not only can place the names of the contents on the skillette activity cans, but you can also code the lids as shown below.

Prepare skillettes related to the skills of measuring, graphing, predicting, interpretation of data, and formulating hypotheses.

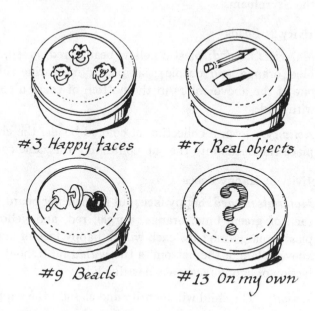

#3 Happy faces   #7 Real objects

#9 Beads   #13 On my own

SAMPLE SKILLETTES

Skillette 1: classifying by color

Objective

> Given a collection of objects, the child will be able to identify and classify them according to color by placing them in the appropriate container.

Activity 1: Toothpicks

> *Materials needed:* plastic margarine tub; colored toothpicks (blue, red, yellow, green, purple, orange, and white); a circular piece of Styrofoam containing circles of the same colors on top.

> *Activity:* Given a collection of colored toothpicks, the child will stick the toothpicks into the circles of matching color on the Styrofoam.

Activity 2: Beads

> *Materials needed:* plastic hollow beads (red, green, yellow, blue, orange, and purple); six plastic margarine tubs with one of the above colors in the bottom of each; a container with a lid.

> *Activity:* Given a collection of colored beads, the child will place the beads into the tub containing the same color.

Activity 3: Happy Faces

> *Materials needed:* happy faces cut from poster board (twelve each of green, blue, orange, purple, red, and yellow); six plastic margarine tubs each with a happy face of one of the above colors in the bottom; a tub with a lid to hold the collection of happy faces; a can with a lid.

> *Activity:* The child will identify and classify the happy faces according to their color by placing them in the tubs with happy faces of the corresponding colors on the bottom.

Activity 4: Game

Materials needed: large coffee can and lid; colored tooth-picks; a plastic Baggie, Styrofoam cutouts of a circle, square, rectangle, and triangle, each containing a small circle of six colors (red, blue, green, yellow, orange, and purple).

Activity: Each child will select a shape from the Baggie and take a turn drawing a toothpick from the can without look-ing. Then the child will place it in the matching colored cir-cle. After the child "draws" a toothpick, he or she will iden-tify and name the color. The first one to get all six colors wins the game.

Skillette 2: classifying by shape

Objective

Given a collection of objects, the child will be able to iden-tify and classify them according to their shape by placing them in the appropriate container.

Activity 5: Colored Poster Board

Materials needed: large coffee can and lid; twelve each of posterboard circles, squares, rectangles, and triangles (blue, red, and yellow); four margarine tubs with the shapes on the sides of the tubs.

Activity: Given a collection of colored shapes, the child will identify and classify each according to its shape by match-ing each to the corresponding shape on the side of one of the plastic containers.

Activity 6: Twister Seals

Materials needed: coffee can and lid; forty-eight candy-striped twister seals formed into circles, triangles, rectan-gles, and squares; four small plastic containers marked with the different shapes; and a red cheese container and lid to hold the shapes.

Activity: The child will identify and classify the twister-seal shapes by placing each one into the plastic container marked with the same shape.

Activity 7: Real Objects

Materials needed: three small paper plates marked with a blue circle, square, or rectangle; a large coffee can and lid; and a collection of real objects. For square objects, use napkins, gauze, wrapped caramel candy, sponges, stamps, fabric, or pillboxes. For rectangular objects, use gum, matchboxes, rulers, Band-Aids, soap, fabric, index cards, raisin boxes, stamps, cough drop boxes, name cards, or envelopes. For circular objects, use buttons, candy mints, bottle caps, curtain rings, pennies, lids, notebook rings, washers, powder containers, Frisbees, or gummed reinforcers.

Activity: Given a collection of real objects, the child will identify the shape and classify the objects by placing them on the plate marked with the same shape.

Related Activity: The child could find other objects either at home or at school and classify them in the same manner.

Skillette 3: classifying by size

Objective

Given a collection or pairs of similar objects of different size, the child will be able to identify the ones that are larger or smaller and classify them by placing them in the appropriate container.

Activity 8: Felt

Materials needed: large coffee can and lid; a piece of black felt (twelve by eight inches); red yarn (eighteen inches); two envelopes; large and small pieces of felt cut into the shapes of large and small oranges, apples, bananas, grapes, and pears; five large and small felt circles, squares, rectangles, ovals, and triangles in different colors.

Activity: Separating the black felt into two sections with the red yarn, the child will identify and classify the felt objects according to size by placing the larger objects on one side of the yarn and the smaller objects on the other side.

Activity 9: Beads

Materials needed: small coffee can and lid; large and small blue and white hollow beads; red cheese container and lid; two plastic containers, one marked with a large piece of blue contact paper, and the other marked with a small piece.

Activity: Given a collection of large and small beads, the child will identify them as large or small, and classify them according to size by placing the larger beads in the container marked with the larger piece of contact paper and the smaller ones in the container marked with the smaller piece of contact paper.

Activity 10: Real Objects

Materials needed: large coffee can and lid; twelve pairs of similar large and small objects, such as spools, corks, crayons, yarn, nails, rocks, paper clips, notebook rings, buttons, safety pins, rubber bands, and shells; large margarine tub marked with a large blue symbol and a small tub marked with a small blue symbol.

Activity: Given a variety of pairs of large and small objects, the child will classify them according to size by placing the larger objects in the large tub and the smaller objects in the small tub.

Skillette 4: classifying by density

Objective

Given a collection of objects, the child will be able to classify them by density by placing each object in water and determining which objects float and which do not, then placing each object in the appropriate container.

Activity 11: Real Objects

Materials needed: large coffee can and lid; large plastic container for the water; three margarine tubs — one marked with a blue arrow pointing up, one marked with a blue arrow pointing down, and an unmarked one to hold the objects; sponge to wipe up spills; paper toweling to dry off the objects; variety of objects such as a cork, marble, rock, shell, straw, rubber band, button, penny, plastic toy, candle, crayon, acorn top, paper clip, nail, notebook ring, Styrofoam, sponge, foil ball, twister seal, feather, nylon net, and plastic disc.

Activity: Given a collection of objects, the child will classify them by density by placing each object in the container of water, one at a time, and put the ones that float in the container marked with the arrow pointing up and the ones that sink in the container marked with the arrow pointing down.

Skillette 5: classifying by texture

Objective

Given a collection of objects, the child will be able to identify by feeling and observing which objects are rough and which are smooth. He will classify them according to rough or smooth texture by placing them in the appropriate containers.

Activity 12: Rough and Smooth Objects

Materials needed: coffee can and lid; three plastic tubs, one to hold the objects, one marked with a piece of rough burlap, one marked with a piece of smooth contact paper; rough objects such as sandpaper, walnut shell, acorn top, burlap fabric, nylon net, sponge, screw, hickory nut, rock, pine cone; smooth objects such as ribbon, facial tissue, marble, satin, smooth rock, acorn, foil, cardboard, tile, waxed paper, plastic rod, drinking straw.

Activity: Given a collection of objects, the child will determine, by feeling and observing, which objects are rough and smooth. He will classify them as to rough or smooth texture by placing the rough objects in the container marked with the burlap and the smooth objects into the container marked with the smooth paper.

Skillette 6: classifying on my own

Objective

Given an opportunity to select objects, each child will be able to classify them according to the property that interests him the most.

Activity 13: On My Own

Materials needed: large coffee can and lid; six plastic margarine tubs; objects selected by the children.

Activity: The children will select their own objects to classify according to the properties that they are most interested in. Some may wish to use some of the same objects that are included in the module and some may wish to provide their own. They will place the objects in the appropriate containers, which could be marked with contact paper and changed as needed.

PA 4
THE THREE-MICE DIET

You have three cages of white mice. To the mice in one cage you give a fortified diet; to those in another cage you give an average diet; and to the remaining mice you give an inadequate diet.

After several months of this kind of treatment you discover:

— that the mice getting the fortified diet are healthy and strong;

— that the mice getting the average diet appear to be healthy and strong, but not quite as vigorous as the mice who received the fortified diet;

— that the mice getting the inadequate diet also seem healthy and strong and, surprisingly enough, more vigorous than the mice who received either the fortified or the average diet.

What can you infer from these results? Are these the results you expected? What explanation can you suggest? Can you suggest any hypotheses that can be tested?

SCIENCE-BOOK KITS

Science-book kits are designed to supplement or replace basal reading textbooks. Each kit includes not only the appropriate book but also the necessary accessories to go with it, including:

— cassette tapes,
— filmstrips,
— records,
— flannel board shapes,
— task cards,
— one large box to hold all the materials (including the book).

We are suggesting science-related titles, but any book will work as long as you are willing to include the necessary ancillary materials. Each book kit should include the following components, contained in individual envelopes when deemed appropriate:

— reading assignments and discussion questions printed on 4 x 6 inch index cards. These questions can be discussed with a friend, teacher, or in small groups.
— spelling words taken from the book.
— activities using the spelling words that are designed to increase competence in pronunciation of words and in understanding of meanings.
— vocabulary words that are introduced in the book.
— vocabulary activities that are structured to provide help with word meanings.
— an evaluation question that is thought provoking and requires a creative answer.
— a culminating task-card activity.
— a manila folder for each child who wishes to use the book kit.
— a supplemental list of books related to the book-kit theme.

We have found that book kits can be used to reinforce and develop the same skills found in basal readers. Also, book kits allow the child to increase reading competence while having freedom of selection.

It takes approximately two to four weeks for a child to complete a book kit, working about one to one-and-one-half hours per school day.

The greatest cost is in buying the books; however, many books are already in the school library.

As a specific example, let us use an intermediate-level paperback book, Scott O'Dell's *Island of the Blue Dolphins* (Boston: Houghton Mifflin, 1960).

Into a large container such as a dress box or cardboard box you will need to place:

1. all media materials related to this paperback book. (You *will* need to read it first!) Try locating records, films, filmstrips, and tapes for starters.
2. a list of discussion questions relating to the book, placed on index cards.
   — Why do you think that Ramo thought of the red ship as a red whale?
   — What are toyon bushes?
3. a list of spelling words from the book, such as:
   — *cricket, carcass, horizon, utensils, anchor, punish*
4. a list of spelling activities to go with the words:
   — A crossword puzzle using at least ten of the new words.
   — Unscramble the message in this code. Some clues are provided:
   VJG (The) FQNRJKP RWTUWGB DA (by)
   VJG YCTTKQTU KP (in) ECPQGU OCPCIGCQ
   VQ HNGG HTQO VJKU FCPIGT
5. a list of vocabulary words from the book, with sentences showing the context in which they're used:
   — *toyon bush.* "The *toyon bush* shielded Karana."
   — *kelp.* "The *kelp* beds were clearly seen from the shore."
6. vocabulary activities using the new words:
   — Select one of your vocabulary words and find several pictures to illustrate its meaning.
   — Take your vocabulary words and place them into several categories. Try colors, nouns or verbs, and other topics of your own choosing.
7. thought-provoking evaluation questions:
   — What do you think was the most exciting episode in this book and why do you feel this way?
   — Name three different emotions Karana felt while on the

island. Give three incidents that would show each of these emotions.

— Explain your feelings about this book.

8. several culminating task-card activities:

— Prepare a picture book of the animals mentioned in this book. Write a short paragraph about each of them.

— This particular Indian tribe was superstitious. Find several superstitions and illustrate them. How did your favorite superstition get started?

— Find four or five poems about the sea and share them with a friend.

— Prepare your own map of the *Island of the Blue Dolphins*. Include the landmarks mentioned in the book. Be sure to include a map key.

— Find out about other islands. What is an island, anyway? Look up Hawaii and Easter Island or countries like Japan and Britain.

9. a supplemental book list for those children who wish to read additional books related to the subject. Try finding dolphin books, island books and/or such books as *Tinkerbelle, Kon Tiki, Swiss Family Robinson*; and magazines such as *National Geographic, Ranger Rick,* or *National Wildlife*. The list can be limitless.

Why don't you try to prepare several book kits and introduce them to the children? Find out if they like them and want more.

PA 5
THE TIGER IS MISSING!

A zoo keeper discovers one morning that the tiger is missing. The door to the cage is open. The door appears not to have been forced open. There are tire tracks leading up to the cage. A man's footprints are leading in and out of the cage, but there are no tiger footprints outside the cage. There are some blood stains on the cage door and some pieces of meat left in the cage.

What might you guess happened to the tiger?

REINFORCE SKILLS WITH THE BUZZ BOX

As we begin to personalize our instruction in science as well as in all aspects of the curriculum, we need aids that assist the teacher in handling the more routine duties so that he or she can concentrate on designing the child's learning experience. With the trend toward student involvement in learning rather than emphasis on a single-textbook, group approach, aids are needed to involve each child. The buzz box is one solution to the problem relating to the reinforcement of individual skills.

The buzz box is a simple battery-operated device that buzzes when a child chooses a correct answer on a teacher-made question card (see the figures). The punched cards are made from manila cardboard cut to 3 x

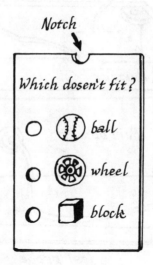

5½ inches. In most cases, three choices are printed on the card. The child takes the punch card to the box, places the punch card on the box top, and selects one of the three choices with the test probe. If this selection results in a buzzing sound, he or she knows his answer is correct. This setup provides the child with immediate reinforcement if the answer is correct, and challenges him or her to try again if it is not.

Adapted from Gerald H. Krockover and Diane Bobb, "Buzz-Boxes — An Aid to Individualizing Instruction," *Science Activities*, vol. 7, no. 2 (April 1972), pp. 26–29.

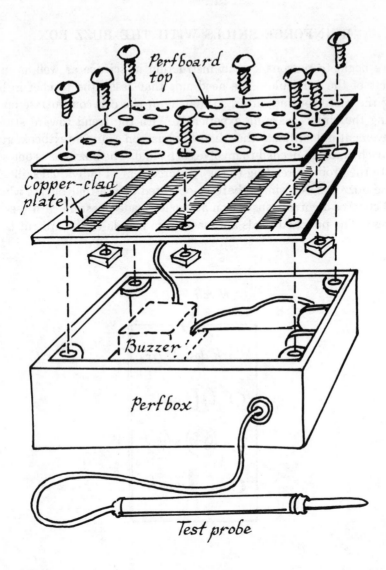

Perfboard top

Copper-clad plate

Buzzer

Perfbox

Test probe

On the classification card illustrated in the figure on the preceding page, the child should select the square block because it is not round. A buzzing sound would inform the child that the answer is correct. If the child does not hear a buzz, he or she will know to try again. The classification possibilities are endless. Color, shape, toys, pictures of animals, chemical formulas — all can be readily adapted to the buzz box. It can also be used for nomenclature, chemical cycles, classification of plants and animals in biology, symbol recognition, and calculations in mathematics and physics. Buzz boxes are also excellent for use in rock and mineral identification in earth science.

To date, our preservice and inservice teachers have successfully constructed more than 200 buzz boxes for use in all curricular areas from the preschool level through elementary, secondary, and college levels.

If you would like to make a buzz box or buzz boxes, a complete step-by-step procedure as well as drawings and a parts list can be found in the "Shoestring Sciencing" section of *Creative Sciencing: Ideas and Activities for Teachers and Children*. All materials are available from an electronics store, and the cost is minimal.

We have found that children above third or fourth grades enjoy building buzz boxes and preparing buzz-box cards. The noise is as small as the cost, but the educational possibilities are endless. The only limitation is your creativity and imagination. Build a buzz box today!

try it — you'll like it!

Build a buzz box with a light instead of a buzzer. Build a buzz box with switches instead of a probe. Use variations for buzz-box cards including computer cards, old greeting cards, index cards, or old plastic charge cards.

SCIENCE GAMES

Are you ready for Groovy Zoo, Quizmo, Animal Friends, Beat the Path, Who Am I?, Pollution Solution, The Survival Game, Animal Dominoes, Draw Me, Trapped, Glacierama, and Bogy? These are but a sampling of the more than one hundred science games that students and teachers have developed themselves. Games can be used to reinforce skills that have already been introduced, such as graphing and classification. But games can also be used as motivational devices for the introduction of a skill, or even for evaluation purposes. Children enjoy solving problems and being involved in thinking activities by playing games. Science games can also be used to reinforce the skills of socialization and interaction with peers through sharing, taking turns, and verbal interaction. Games can serve as one mechanism for the humanization and relaxation of the classroom atmosphere. Although some games are original, many science games are patterned after popular commercial games. Many require paper and pencil only; others require a game board and game cards.

We will present several examples; you will be able to think of many more. Some of the ideas and rules are explained on the next few pages. Change the idea and rules to fit your needs and the needs of your students.

After you've got the idea, develop a game to reinforce the skill of measuring metrically or the skill of inferring. How could our puzzler activities be grouped together and game rules prepared for them?

SCIENCE GAMES

1. Groovy Zoo (a classification game, invented by three third-grade boys!)

   You will need poster board to make a playing board with a start and a finish and divided into small units, each unit naming a typical animal characteristic; four or five model clay animals, such as a fish, rabbit, turtle, and a snake, to be used as counters by the players; and a numbered spinner.

   Each player takes one of the clay animals as a counter and places it on the playing board at the starting point. Each player spins a turn in the regular playing order. The number a player spins is the number of spaces he or she may advance *if* the clay animal has the characteristic named on the unit the player will land on; if the animal doesn't have that characteristic, then the player must stay where he or she is and wait for another turn. The first player to reach the finish is the winner.

2. Quizmo (an observation game)

   You will need to make a set of animal description cards. Each card will carry a good description of a certain animal but will not name that animal. You will also need a different set of players' cards. Player cards are pieces of manila board divided into sixteen squares. Each square carries the name of an animal that is described in the set of description cards.

   One person is the "caller" for the game. The caller draws a description card from the deck and reads it; the player covers the name of the animal described. The first person to correctly cover four names in a horizontal, vertical, or diagonal row is the winner.

3. Who Am I? (an observation game)

This game uses the playing board and spinner used in "Groovy Zoo." But for each characteristic named on the front of the board, you will have to list the animals that have that characteristic on a separate sheet. No more than four should play this game at one time.

The players decide on a playing order, and then they spin in turn. Each player moves the number of spaces indicated by the spinner. When the player lands on a space, he or she must read the characteristic named there and then name an animal that has it. If the player cannot, he or she must go back to where he or she began the turn. Answers can be checked by referring to the answer sheet. The first one to reach the finish is the winner.

4. The Animal Game (an observation game)

You will need a game board, a deck of seventy-two cards, and two or more players. Your game board should have the pictures of about one hundred different animals on it. (You may also want to number the pictures and make a keyed list of the animals' names.) Each card in the deck should name one animal characteristic. Limit the total number of characteristics you use to about thirty, and use them twice or three times on different cards.

The players are each dealt nine cards. The remainder of the deck is placed, face down, by the board, and the top card is placed face up beside the deck. Each player, in turn, takes the upturned card or the top card of the deck and adds it to his hand; then he may match a card with one of the different animals, or he may discard that or another card onto the face-up card. When a player accumulates three cards that describe one of the animals on the board, the player may place those cards over the animal's picture (but only in turn!). The first player to "go out" — to get rid of all his or her cards — is the winner.

5. Draw Me (an observation game)

All that's needed for this game is a bit of imagination, an object or a picture, paper, and pencils. Pass out the pencils and paper to all the players. Then, holding the object or picture so that no one can see it, describe it *without naming any of its parts*. Your description should be limited to how the object "looks" — its size, shape, color, texture, and so forth. The object of the game is to see who — if anyone — can correctly draw the object you are describing. Sound too easy? That's only because you haven't tried it!

6. Animal Dominoes (an observation game)

You'll need to make a set of at least twenty-eight animal domino cards. Each card should be divided into equal halves, with each half carrying a picture of a mammal, bird, fish, amphibian, reptile, or insect (this is their relative value, in descending order). The domino cards should carry every possible combination of pictures — double mammals, mammal and bird, mammal and reptile, bird and insect, and so on.

Seven of the cards should be dealt, face down, to each of the two to four players. The player with the highest double mammals card begins the game by placing the double face up at the center of the table. The player to his or her left must play a match or draw another card from the "bone pile" (the leftover cards). If the player still cannot play a match, he or she loses the turn. The players continue to match or draw until one player has matched all his or her cards: that player is the winner.

MODULES

An alternative creative-sciencing technique is that of modular instruction. Before we examine several actual instructional modules, let's consider the components of a module and its advantages and disadvantages. Modules are also sometimes called learning-activity packages, unit packages, teaching-learning units, minicourses, mods, individualized learning experiences, and so forth. Keep in mind that only the name changes, the idea is the same. We will use the general term *module* to refer to any type of learning package that is designed to involve children on an individual basis. Our examples will be science examples; however, you now realize that any curricular subject can be used, as can any theme approach (we will consider themes in the section on learning centers, below). We have utilized modules at all levels — preschool through adult — and have found them to be an effective learning tool.

What is a module?

A module is an instructional unit that can stand on its own, and it is based on a single concept or area. A module is designed to assist the student in the attainment of a specified objective or instructional goal. All materials and activities needed to attain this instructional goal are included in the module. Those teachers wishing to individualize their instruction should design a preassessment evaluation to use with their instructional modules. As a result, only those students requiring a specific goal are required to become involved in a specific module. In the same manner, a postassessment evaluation allows the student and teacher to determine whether or not the module assisted the student in becoming competent in a specific goal or objective.

Components of a module

A module usually contains the following components:

- the title of the module, plus any necessary background information;
- the instructional objectives for the module (usually no more than two to four since we are dealing with packets of instruction);
- a preassessment evaluation to determine if the learner needs the module for competency;

— a list of all materials needed for the student to perform the modular activities;

— the instructional procedures, which include any necessary readings, laboratory activities, or investigations;

— a generalizing experience (making the specific module applicable to other topics and areas);

— a postassessment evaluation. We prefer to use a generalizing experience as a postassessment evaluation; however, you may wish to prepare a separate postassessment evaluation.

Advantages and disadvantages of modules

The advantages and disadvantages of modular instruction are continually debated. As we see it, here are the advantages (strengths) and corresponding disadvantages (weaknesses) of modular instruction:

Advantage: Provisions can be made for the assessment of each individual child.

Disadvantage: Sometimes individualization takes place at the expense of the entire class.

Advantage: Modules reduce the amount of time the teacher lectures to the whole class and increase the personal contacts between teachers and students.

Disadvantage: The teacher sometimes finds himself or herself spending too much time answering procedural questions.

Advantage: Modules allow the teacher to spend less time covering material that the student already knows.

Disadvantage: Sometimes group review can be beneficial for all students.

Advantage: The modular approach is self-instructional, with the child proceeding at his or her own rate or pace.

Disadvantage: Not all children can discover or work at their own pace and find the experience satisfactory.

Advantage: Modules require fewer materials and less equipment, since not all students are doing the same thing at the same time.

Disadvantage: There may be twenty-five different things going on at the same time in the classroom.

Advantage: Modules allow increased individual involvement in one's own instruction.

Disadvantage: In this way, group interaction is reduced.

Advantage: Since modules are small segments of learning experiences, they can be easily updated.

Disadvantages: Modules take too much teacher involvement and time after school.

Are modules worth the effort? We suggest that you try some with your students and then you be the judge. You may wish to duplicate the examples presented here or try writing some of your own. We cannot decide if modules are for you; only you can make that decision.

As we prepare our own instructional modules, we find that it helps to ask ourselves five questions. Try asking yourself these questions as you prepare your own modules:

1. What does the student need to know about this idea, concept, or skill? (Or, what are my specific instructional *objectives?*)
2. Does the student already know it? (What form should the *preassessment evaluation* take?)
3. What *materials* will I need to provide for the student to achieve the objectives?
4. How will I do it, and what will the student be doing? (What is the *instructional procedure?*)
5. How will the student and I know that he or she has arrived? (What would be a good *generalizing or postassessment experience?*)

MODULE 1
CLASSIFYING COLORS AND SHAPES

Level: primary grades, or K–3

Objectives: At the end of this module the student should be able to:

— identify each of the following shapes: circle, triangle, and square;
— identify each of the following colors: red, blue, green, yellow, orange, purple, black, and brown;
— identify two-concept objects: red circle, blue square, yellow triangle.

Preassessment evaluation: First, the child is shown color cards and is asked to identify what color the cards are. Then, the child is shown cardboard shapes and is asked to identify them. Finally, the child is shown colored shapes (such as a red triangle) and is asked to identify them by color and shape.

   If the child does not successfully complete each preassessment task (does not score 90%), then he or she should do the module.

Materials needed: three shoe boxes; wood or plastic building blocks of various colors; a magnet; wires shaped as circles, triangles, and squares; and cardboard squares, triangles, and circles that are yellow and green, as well as the original cardboard pieces from which they were cut.

Instructional procedure: Prepare three activity boxes that can be given to a child for instructional purposes, their contents as indicated below:

1. A construction box, with all the colored building pieces.
   Ask the child to build a yellow building or a red building.

2. A magnet box, with a magnet and the wire shapes of circles, squares, and triangles. Ask the child to pick up a square, circle, or triangle.
3. A puzzle box, with the yellow and green cardboard cutouts of squares, triangles, and circles. Ask the child to put a green circle in the circular hole, or a yellow triangle in the triangular hole.

Generalizing experience: Give the child a piece of paper with the outlines of a circle, square, and triangle on it. Also provide the child with a box of crayons with the eight basic colors. Then ask her or him to find the circle and color it red.

Continue this procedure using a variety of the possible combinations until the skill has been satisfactorily reinforced.

MODULE 2
CONTROLLING AND MANIPULATING
VARIABLES IN CAPILLARITY

Level: intermediate grades, or 4–6

Objectives: At the end of this module the student should be able to:

— identify and name five variables which are important in influencing either the rate at which the water moves or the maximum height to which it rises;
— construct an investigation in which he or she holds all but one variable constant;
— identify the dependent and the independent variables in the investigation.

Preassessment evaluation: Give the student a shallow plastic dish or cup containing water about three centimeters deep. Give him or her three-by-ten centimeter strips of two different colored materials, such as paper towel or cotton. Ask the student to show that the liquids move at different rates in different materials by constructing an investigation that will allow you to observe this phenomenon. Have the student name the independent and dependent variables. The student will not need to proceed if:

— he or she can construct an investigation in which he or she holds all but one of the variables constant;
— he or she can name the independent and dependent variables in his or her investigation;
— he or she can name five variables which influence the rate at which the water moves or the maximum height to which it rises.

Materials needed: three or four different types of material, four or five plastic dishes or cups, clock with a second hand, liquid, meter stick.

Instructional procedure: Using a strip of blotting paper and a shallow dish with water, the student observes water moving easily up the blotting paper. After this action has been observed, the student should prepare a list of inferences that suggest what variables are important in influencing either the rate at which the water moves or the maximum height to which it will rise. Then, he or she should design an investigation in which all but one of the variables are held constant while only the independent variables are manipulated. The interpretation of students' observations will be a sure test of their inferences. Typical observations that students will make are:

— The height of the water is almost the same on any two strips of blotting paper used.
— The water moves up the strip.
— The water moves up the two strips at about the same rate.

The student should then try to determine why the height of water was the same for the two strips. Students will need to consider the variables in these questions:

— Were both strips the same size and shape?
— Were both strips put the same distance into the water?
— Were both strips in the water for the same length of time?
— Were both strips made of the same material?
— Did both containers hold the same liquid?
— Was the temperature of the water in both containers the same?
— Would it make a difference if both strips of paper were held by a different person?

— Does color of material influence the rate at which the liquid rises?

The student should see the need for keeping all but one of these variables constant in the investigation.

The student should now suggest questions to consider for the next portion of this investigation. Sample questions might include the following:

— Would the waterline be the same if the two strips of paper differed in width?
— How does the rate at which the water rises depend on the width of the blotting paper?
— Would the waterline be the same if the two strips of paper were dipped in the water to different depths?
— Does the rate at which the water rises depend on the type of material that the strips are made of?
— Would the waterline be the same if the two strips of paper were left in the water for different amounts of time?
— Does the height of the waterline depend on the temperature of the water?
— Would the height of the waterline be the same if two colors of the same type of paper were used?

The student should design an investigation to test some of the questions he or she has proposed. He or she should keep in mind that all variables but one must be kept constant. The student should be able to identify which variable is the dependent variable and which variable is the independent variable.

Generalizing experience: The student should be given two plastic containers and two identical strips of a material different from those used before. Put water in one container, and put cooking oil in the other to the same depth. Ask the student to investigate the question: Do different liquids move at different rates in the same material? Ask the student to prepare a list of variables he or she has identified. Then ask the student to select one variable from this list and conduct an investigation to try and answer the questions raised by the isolated variable. The student should be able to name the dependent and independent variables in his or her investigation.

If the student does not successfully complete the module he or she should try again, using a different liquid and strips of a new material.

MODULE 3
CONTROLLING AND MANIPULATING
VARIATIONS IN EVAPORATION

Level: middle school, or 7–9

Objectives: At the end of this module the student should be able to:

— identify and name five variables which are important in influencing the rate at which water evaporates;
— construct an investigation in which he or she holds all but one variable constant;
— identify the dependent and the independent variables in the investigation.

Preassessment evaluation: Give the student a balance with two sponge cubes of identical size to hang on each side of the balance. Ask the student to prepare a list of factors that would affect the evaporation rate. Have the student prepare an investigation that will allow her or him to check any two of these factors. Have the student name the independent and dependent variables in the investigation. The student will *not* need to proceed if:

— he or she can construct an investigation in which he or she holds all but one of the variables constant;
— he or she can name the independent and dependent variables in the investigation;
— he or she can name five variables that are important in influencing the rate at which water evaporates.

Materials needed: a single-beam balance, sponge cubes, plastic sandwich bags, light bulb with reflector, water, hot and cold.

Instructional procedure: The student should attempt to think of factors that would influence the evaporation rate of water from the sponge cubes. He or she should attempt to rank these factors in order of importance, as he or she sees it. He or she should then attempt to design an investigation in which all but one of the variables are held constant. The student should manipulate one of the variables while holding the other ones constant. A list of typical variations that could influence evaporation and be tested in an investigation follows:

— Shine a light source on one sponge.
— Use hot water on one sponge and cold water on another.
— Put a closed plastic bag around one sponge.
— Shine a light on both sponges, but protect one with a thin transparent sheet.
— Use different-sized or different-shaped sponges of the same mass.
— Fan air past one sponge.
— Cut one sponge with ragged edges.

The student should attempt to identify which factor is most important in the evaporation process. A good way to identify this factor is to time how long it takes the balance to become unbalanced when each modification is considered. The student should also attempt to relate the experiments to nature — that is, determine the effect of the sun, wind, and surface area upon evaporation.

Generalizing experience: The student should be given a container of water and asked to list those factors he or she would consider as explanation for increased water evaporation. He or she should then design an investigation to determine the evaporation rates for this method. He or she should be able to name the dependent and independent variables for the investigation.

If the student does not successfully complete the module, he or she should return to the instructional procedure and, using a different liquid, conduct another investigation.

THEORY INTO ACTION

We have presented three examples of science modules, developed and used for kindergarten through the middle grades and junior high. These modules are not perfect. They will need to be continually revised to keep pace with student comments and teacher observations.

Modules emphasize science skills and science concepts, not the memorization and regurgitation of facts, facts, and more facts. The emphasis in the primary grades is on the basic skills of science, including observing, inferring, measuring, and predicting. The emphasis in the intermediate grades is on the integrated science skills, such as formulating models, interpreting data, and controlling and manipulating variables. Skills are as valuable as content. You can always look up an answer pertaining to content. Try looking up a skill!

During the middle, junior, and high school years, emphasis should be placed on the further development of the skills previously mentioned. In addition, emphasis should be placed on a greater sophistication in the use of laboratory materials for investigations.

How do you begin to develop modules for your science students?

To begin module implementation, you might start by preparing a core of modules for your science program at a minimal level of student performance. Each child entering should receive an evaluation related to your instructional goals so that a profile of each student can be prepared revealing his or her strengths and weaknesses. In this way, you, as the teacher, will know where this individual student is in relation to the performance outcomes set by you, or you and the student. Once this evaluation has been accomplished, a list of resources should be developed. The national curricular programs in science as well as this book's companion volume can serve as good starting points. In time, a portfolio of activities of a core and supplemental nature will evolve, allowing you to provide a sequence of instruction for each student, taking him or her from where he or she is to where he or she should be. Undoubtedly, on the first time around, these modules will be difficult to develop. It will take time, and a reduction in an individual teacher's work load may be needed to accomplish real development. Teachers who take the time to personalize their instruction and prepare modules of instruction, however, have their directions and goals in mind for their science program. Unless we take the time to do this, we will remain on that never-ending treadmill of vague science instruction. The role of the teacher must shift from one of a custodian of education to that of an architect and a creator of the instructional program in science.

We have reviewed many skills and techniques, including (just to refresh your memory):

— task cards;
— the minipreparation of audiotutorial tapes;
— invitations to investigate;
— skillettes;
— science-book kits;
— buzz boxes;
— science games;
— puzzler activities;
— modules.

Many teachers ask us if mechanisms are available to incorporate all these alternative creative-sciencing endeavors into one concerted teaching effort. The answer is a resounding

YES!

on to interest centers!!

INTEREST CENTERS

Since interest centers are flexible and build on the alternatives previously mentioned, they attract children. With the trend to the integrated curriculum, open-concept learning, and individualized instruction, interest centers are right on target. Interest centers can usually be divided into three categories: content, skill, and theme. Let's examine each type.

Probably the most traditional and easiest type to assemble is a content interest center. It is based on a particular area of content, such as magnets, plants, or electricity. Thus, in a magnets interest center, all activities (task cards, modules, games) would be tied to the content area of magnets.

The second type is the skill interest center. Its sole purpose is the development and reinforcement of a particular skill, such as the skill of classifying or graphing. All activities (task cards, modules, games) would be tied to the skill area of graphing or classifying.

The third type (the most forward-looking and true to life) is the theme interest center. This center is based on broad themes (clothing, cooking, sewing, the future, the circus, communication, the arts, cities, farming, environment). Within the theme interest center we find content and skills drawn from a variety of subject areas and integrated. Since they are so comprehensive, theme interest centers require imagination and creativity of a teacher.

An interest center may be placed in any strategic location in your room, area, or school. An increasing number of teachers are organizing their instruction based on a great variety of interest centers available to their students throughout the school day. Some teachers and some schools are organized completely around the interest-center approach. As one teacher remarked to us: "Interest centers allow me to be with my students. I can communicate with them on a personal basis and I can encourage group discussions. I wish I had learned how to make an interest center in college."

Making an interest center

The first step in making an interest center is developing the related teacher- and student-created materials. These materials include the appropriate teacher guides, student textbooks, reference books, activity sheets, task cards, modular activities, book kits, skillettes, invitations to investigate, puzzler activities, films, tapes, games, and buzz boxes.

All of these materials should be arranged in a sequence based on your instructional goal(s) for the center. Many teachers use a preassessment evaluation to determine which students need which portions of the center. Remember to include evaluation materials related to the interest center's purpose (content quizzes, tests, competency measures, and also attitudinal measures).

Display the materials in an attractive, eye-appealing manner. Packing boxes should be obtained for storing the interest centers when they are not in use.

We have shown a sample theme interest center in greater detail on the next three pages. This example is not intended to be comprehensive. It is a summary of the types of ideas that can be incorporated into one center. Take a look at it and you'll be on the road to a "filling" interest center. Yours will probably be good enough to eat!

THE INTEREST CENTER

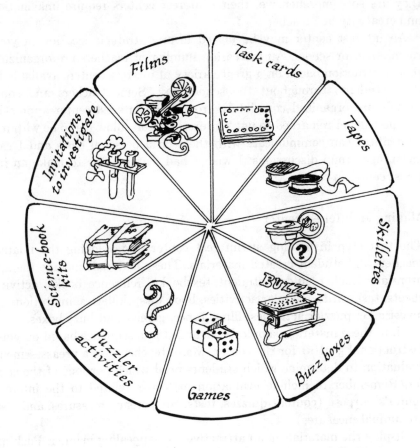

Films

Task cards

Invitations to investigate

Tapes

Skillettes

Science-book kits

Buzz boxes

Puzzler activities

Games

THEME INTEREST CENTER ON COOKING

Task cards: Find your favorite recipe and prepare it to serve to the class.

Media: Find films, slides, transparencies, tapes, pictures, and other materials related to the theme of cooking. Prepare activities to go with them.

Audiotutorial tapes: To teach children to follow directions, prepare several tapes dealing with recipes and have children follow them. Try popcorn, muffins, Rice Krispy cookies.

Science-book kits: Good starters for a science-book kit are Vicky Cobb's *Science Experiments You Can Eat* (New York: J. B. Lippincott, 1972) and *More Science Experiments You Can Eat* (New York: J. B. Lippincott, 1979). Go to the library; you'll find many more ideas for book kits.

Skillettes: Prepare skillettes relating to such skills as classifying, following directions, computing with fractions, or using measuring containers.

Buzz boxes: Prepare buzz-box cards related to the spelling of certain foods or matching food names with their pictures.

Games: Invent several food games patterned after existing games. Develop:

— *Concentration* by matching food pictures;
— *What's My Line* by asking, "What's my cooking-related occupation? Is it butcher, baker, butler, or cookbook writer?"

What about the *Food Chain Game,* or *Who Eats Whom?*

More ideas

— Field trips to the grocery store, factory, restaurant, bakery, home kitchen
— Foods, lands, customs, and people
— Your own class recipe book
— Food mobiles and centerpieces
— New food inventions
— Food cost investigations
— Spoon, fork, knife — Where did they come from and why do we use them?
— Investigate diets and your health.
— What foods will make you healthy?

Invitations to investigate: There are all sorts of possibilities here. Try something like the following:

— Why does dough rise?
— What can you learn about mold? What does it have to do with food?

Puzzler activities: See PA 6 on the next page.

PA 6
WHAT'S COOKING?

You live in a totally electric house with modern appliances in the kitchen. Because of a severe electrical storm last night, you will be without electricity for the next three days. You had planned to have five guests for dinner during this period. In spite of the electrical blackout, you decide to go ahead with your plans. You have a propane camp stove, an ice chest, a box of matches, and water. What will you prepare for dinner? Why?

Since the electric can opener, blender, and electric stove won't work, will you have to change your meal plans? What other appliances will be affected by the loss of electrical service for three days?

What provisions will you make for washing the dishes after the meal?

We have devoted considerable time to describing the alternative skills and techniques you will need to prepare your students for tomorrow. But what about the whole concept of evaluation based on this approach to learning? Are there any models we can use? Can we evaluate children in the affective and psychomotor domains as well as in the cognitive? What about criterion-referenced testing? Section 6 is for all evaluators! But before we begin it, let's explore several examples of an integrated approach to learning in the next section.

PAUSE FOR A SUMMARY

— Creative-sciencing alternative skills and techniques include task cards, puzzler activities, audiotutorial tapes, invitations to investigate, skillettes, science-book kits, buzz boxes, games, modules, and interest centers.

— Task cards (sometimes called idea cards, suggestion cards, or activity cards) are an excellent way to involve children in individualized instruction.

— Puzzler Activities are brief confrontations with real or fictitious situations supplied by the teacher. They challenge the student to map out strategies, raise appropriate questions, and consider different explanations or solutions.

— The minipreparation of audiotutorial tapes (Mini-PATT) allows teachers to produce instructional tapes suitable for their own classes in a short period of time.

— Invitations to investigate are problem-solving situations that allow students to analyze and synthesize an experimental observation.

— Skillettes are short, discrete lessons designed to introduce or reinforce a single skill. The science-process skills can serve as the basis for skillettes.

— Science-book kits are designed to supplement or replace basal reading textbooks. Each kit includes not only the appropriate book, but also all of the necessary accessories.

— The buzz box is a simple battery-operated device that buzzes when a child chooses a correct answer on a teacher-made question card.

— Science games can be used to reinforce skills, to motivate students, or to evaluate progress. Games serve to relax the classroom atmosphere.

— A module is an instructional unit that can stand on its own and is based on a single concept or area.
— Interest centers usually can be divided into three categories: content interest centers, skill interest centers, and theme interest centers.

Science in the integrated curriculum

Great risks, greater profits

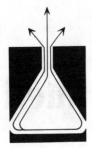

A child's mind is not a sponge. It is more a well-spring.
— *Alfred DeVito*

Science, as a discrete topic, is not the most frequently taught subject in the elementary school. In fact, in many schools it is the least taught subject. It is taught less often at the primary grades than at the intermediate grades. When surveyed, numerous elementary school teachers expressed a strong negative attitude toward science. This disaffection runs the gamut from boredom and dislike to apprehension and fear of science. Paradoxically, these same teachers think science would be the most interesting subject their students might engage in. Yet the battle is lost as elementary teachers, at the expense of science instruction, devote a disproportionate amount of time to other areas of the curriculum. How then is the battle to be won?

Will adding more science content to an undergraduate elementary education curriculum do it? We think not. Will improving the science-methods course do it? Definitely this would help. But we think the successful implementation of science instruction at the elementary level can be accomplished in several ways — by teaching science as a discrete topic, or by teaching science as it weaves throughout the entire elementary curriculum. Teachers must not continue to accept the so-called vaccination theory of education, that says:

English is not History and History is not Science and Science is not Art and Art is not Music; Art and Music are minor subjects and English, History and Science are major subjects; and a subject is something you "take" and, when you have taken it, you have "had" it, and if you have "had" it, you are immune to it and need not take it again.[1]

Elementary teachers would be remiss if they did not capitalize on science as it appears in reading, mathematics, social studies, art, music, and so forth. The following pages provide examples of where and how science can be integrated into the elementary school curriculum.

SCIENCE AND MATHEMATICS

Mathematics is the communication tool of the scientist. In short, mathematics enables science to be quantified. It permits science investigators to describe the time element of the processes of evaporating, boiling, and dissolving. Through the use of mathematics, scientists can also define direction, distance, and speed. Mass, weight, specific gravity, and other properties of matter become relevant when they can be stated mathematically.

Everyone is endowed to some extent with a feeling of the physical space that surrounds us and the events of nature that take place in that space. The motion of the sun, earth, moon, and other heavenly bodies, the falling of a leaf, the sound of rain and wind, as well as chemical and biological changes are often perceptible. Scientists describe these events as accurately as they can. Although some of their description may be verbal or written, it is their use of mathematics that makes their reporting of the phenomena of nature more precise.

Science, as promoted through inquiry, necessitates the collection of data as facts are generated from a science investigation. The collection of data is subsequently analyzed and interpreted; and the resultant conclusions are drawn by the investigator. The mathematics involved in the collection and interpretation of the data can be simple or sophisticated, depending on the capabilities of the user.

Take a deep breath. How much air did you inhale? What is the capacity of your lungs? What is the relation of the air you inhaled to your total lung capacity? On the average, how many times a minute do you inhale and exhale? How does this change with exercise? Is what you exhale the same substance as what you originally inhaled? What hap-

1. Neil Postman and Charles Weingartner, *Teaching as a Subversive Activity* (New York: Delacorte, 1969), p. 21.

pens to the inhaled material to substantially change it by the time it is exhaled? How long can an individual survive without air? What happens to the carbon dioxide after it is exhaled? Is it possible you are breathing recycled air that once was inhaled and then exhaled by Adam, Isaac Newton, Abraham Lincoln? Can you draw or collect pictures of what lungs look like? Why does a medical doctor place a stethoscope against your back or chest and ask you to cough or take several deep breaths?

Additional mathematics examples include:

1. Measuring activities: What is the length of the room? Width? Measurements of volume and mass are also valuable activities.
2. Graphing activities: Make bar graphs of the heights or weights of the children compared to their ages. Try graphing the growth of a plant.
3. Collecting data from advertisements: Which paper towel is best? Also try going on a litter hunt. Determine how many kinds and what amounts of litter you can find and which type of litter is most common.
4. Puzzler Activities: The ones included in this book provide some examples of story problems that can be related to mathematics.

Creative Sciencing: Ideas and Activities for Teachers and Children, the companion volume to this textbook, contains more than forty activities that integrate mathematics and science.

SCIENCE AND METRICS

The United States is going metric and it behooves each elementary teacher to help prepare children for life in a metric world. Science and mathematics make a natural starting point for metric instruction. Then, all other areas of the curriculum, such as reading, language arts, or social studies, can be included. Tables 1, 2, and 3 have been included to help you and your students become familiar with the metric system.

If you still think that it is easier to use the English (now called Imperial) system than the metric system, try these problems — time yourself on each one (you may use a calculator).

1. Area (Length × Width)
   Problem: What is the area of the floor of a room with the following dimensions:

	Imperial	*Metric*
Length	15 ft 7 in	4.75 m
Width	12 ft 6 in	3.81 m

Answer: _____ ft²   _____ m²

Time to make calculations _____   _____

194¹⁹/₂₄ ft²   18.10 m²

2. Mass (sometimes incorrectly called weight)

   Problem: What is the total mass of the contents of a basket that contains the following items:

	Imperial	*Metric*
Meat	4 lb 9 oz	2.07 kg
Potatoes	3 lb 4 oz	1.47 kg
Tomatoes	2 lb 15 oz	1.34 kg
Cereal	1 lb 7 oz	650 g

Answer: _____ lb ___ oz   _____ kg

Time to make calculations _____   _____

12 lb 3 oz   5.53 kg

3. Volume

   Problem: What is the total volume of the two comparable, but not equal, mixtures?

	Imperial	*Metric*
Milk	1 gal, 2 qt, 1 pt	6.5 l
Water	3 qt, 1 pt	3.5 l
Flavoring	½ pt	250 ml

Answer: _____ pt   _____ ml

Time to make calculations _____   _____

20½ pts   10.250 ml

Total time for Imperial calculations _____

Total time for Metric calculations _____

Now which system do you think is easier to use? Why?

Science and Metrics / 135

remember: think metric!

Do not waste time making children calculate conversions between the two — use tables for this purpose.

To help children relate Celsius temperatures to everyday experiences, share this scale with them:

TABLE 1

CELSIUS TO FAHRENHEIT CONVERSION TABLE

°C	°F	°C	°F	°C	°F	°C	°F
0	32						
1	34	26	79	51	124	76	169
2	36	27	81	52	126	77	171
3	37	28	82	53	127	78	172
4	39	29	84	54	129	79	174
5	41	30	86	55	131	80	176
6	43	31	88	56	133	81	178
7	45	32	90	57	135	82	180
8	46	33	91	58	136	83	181
9	48	34	93	59	138	84	183
10	50	35	95	60	140	85	185
11	52	36	97	61	142	86	187
12	54	37	99	62	144	87	189
13	55	38	100	63	145	88	190
14	57	39	102	64	147	89	192
15	59	40	104	65	149	90	194
16	61	41	106	66	151	91	196
17	63	42	108	67	153	92	198
18	64	43	109	68	154	93	199
19	66	44	111	69	156	94	201
20	68	45	113	70	158	95	203
21	70	46	115	71	160	96	205
22	72	47	117	72	162	97	207
23	73	48	118	73	163	98	208
24	75	49	120	74	165	99	210
25	77	50	122	75	167	100	212

TABLE 2

THE METRIC SYSTEM

Measures of length	10 millimeters (mm) = 1 centimeter (cm)
	10 centimeters = 1 decimeter (dm)
	10 decimeters = 1 meter (m)
	1,000 meters = 1 kilometer (km)
Measures of area	100 square millimeters (mm²) = 1 square centimeter (cm²)
	100 square centimeters = 1 square decimeter (dm²)
	100 square decimeters = 1 square meter (m²)
Measures of volume	1,000 cubic millimeters (mm³) = 1 cubic centimeter (cm³ or cc)
	1,000 cubic centimeters = 1 cubic decimeter (dm³)
	1,000 cubic decimeters = 1 cubic meter (m³)
Measures of liquid volume	1,000 milliliters (ml) = 1 liter (l)
	Note that 1 cubic centimeter of volume is approximately equal to 1 ml of liquid volume; 1 ml of water weighs approximately 1 gram (g).
Measures of mass	1,000 milligrams (mg) = 1 gram
	1,000 grams = 1 kilogram (kg)
	1,000 kilograms = 1 metric ton

TABLE 3

BASIC METRIC UNITS, THEIR ABBREVIATIONS, AND EQUIVALENTS IN IMPERIAL UNITS

Measure	Metric unit	Abbreviation	Equivalent in Imperial measure
Length	Meter	m	39.37 inches
Volume	Liter	l	1.06 quarts
Mass	Gram	g	0.035 ounce
Weight or force	Newton	n	0.224 pound
Temperature	Degree Celsius	°C	1.8 degrees Fahrenheit
Heat	Calorie	c or cal	0.004 British Thermal Unit (BTU)

— flaming forties (35°–45°C): heat wave
— thirsty thirties (25°–35°C): hot summer temperatures
— temperate twenties (15°–25°C): cool to warm
— tingling tens (5°–15°C): cool
— frosty fives (minus 5°–plus 5°C): cold winter weather

In conducting activities, all measurements should be made using the metric system. Linear measurement activities should include having the children measure their bodies, objects in their classroom, and objects at home, and make "longer than" or "shorter than" comparisons. Measurements of mass should first include estimating the masses of objects and then weighing them. The cost per kilogram of various foods should be calculated. Volume and capacity measurement activities include calculating the capacity of a graduated cylinder and then checking the calculation by actually measuring the capacity; and making metric meals using metric recipe cards.

Other activities include a school metric olympics; a "think metric" week at school or in the community; writing metric commercials; providing daily metric weather reports; making a metric road map for your state; writing a metric poem or song; finding careers that use metrics — pharmacists, doctors, photographers, tire and automobile manufacturers. Many other ideas for metric activities may be found in the companion volume, *Creative Sciencing: Ideas and Activities for Teachers and Children.*

Additional metric resources include:

1. American National Metric Council, 1625 Massachusetts Avenue, NW, Washington, DC 20036
2. National Bureau of Standards, Washington, DC 20402
3. Slide Chart Corporation, PO Box 527, West Chester, PA 19380 (for their Metric Slide Chart Calculators)
4. *Activities Handbook for Teaching the Metric System* by Gary G. Bitter, Jerald L. Mikesell, and Kathryn Maurdeff (Boston: Allyn and Bacon, 1976).

points to ponder

Why is the United States the last major country to go completely metric?
    How would you convince parents and other adults that children should learn metrics in school?

How might you include metric questions to encompass music? Art? Social Studies? Reading?

SCIENCE AND ART

The following are some of the infinite possibilities for science-related art activities. These suggestions are categorized according to topics rather than age or grade level. You must be the judge as to their appropriateness and you must be the facilitator in upgrading or simplifying the suggestions to fit your particular needs.

Motion   Have the children paint realistically or abstractly, using brush strokes, lines, forms, and colors that suggest motion. This can be done to music or with a film (for example, a picture of a moving merry-go-round). Relate this to science by stressing forward motion, sideward motion, circular motion, clockwise motion, counterclockwise motion, rotation, revolution, reciprocating motion, and a spiraling volution.

Sensitivity to motion   Have the children graphically describe the differences between the motion of a falling feather and a dropped rock.

Compare, by the particular character of lines, different views of the same subject, running, walking, jogging, catapulting through space. Select several objects that travel through space such as paper airplanes, Frisbees, and whiffle balls. What art would be representative of the flight patterns of these objects?

Drawing of simultaneous motions   Have the children represent simultaneous motions of a spinning object such as a spinning top or a gyroscope. Two kinds of motion may be shown — a radial motion accompanied by a spiraling motion.

Sculpture that moves   A simple mobile or wooden chimes may add a great deal to an activity on the observation of wind and sounds. A lighted candle mounted on a lazy Susan placed in motion should result in some interesting line-trace patterns when depicted by students.

Color fusing   Place two liquid watercolor paints side by side on a sheet of wet watercolor paper. Tilt the paper to obtain varied intermixing. How will the effect vary when equal amounts of separate colors are used?

Unequal amounts? This can rapidly lead to activities related to porosity, permeability, capillarity, and density.

Oil on water   Place a few droplets of oil on the surface of some water. Agitate the oil. Place the water and oil in the sunlight so that the wave-motion dispersion of color may easily be seen. Have the children paint a nonfigurative design creating a similar effect.

Spectrum composition   Water is an excellent reflector and refractor of color spectrum in rays of light as observed in a rainbow sometimes associated with a rainstorm. Soap film on water will act like oil on a puddle or raindrops in a cloud, catching and refracting the colors of the spectrum.

Mixing colors   Isaac Newton invented the color wheel to explain to his students what he had discovered about refracting light through a prism. Have the children mix colors both in a dry and in a wet state. Try using two basic colors and do a finger-painting by intermixing the colors.

How would you extend this work with colors into other art activities that could be associated with science, for example: shadows, light reflected in a mirror, transparency, and translucency?

A total environment of movement   This activity is aimed at having children respond to a total environment of movement rather than a single, one-shot description. What motions would be involved in a carnival, a Saturday high school football game, a busy harbor scene? The honking of horns, the flapping of flags, or the movement of masses of people create a symphony of motion. Try showing these activities in art investigations.

SCIENCE AND OTHER THINGS

Movement in a confined space   Have the children represent the movement of an elevator, or have them draw a spiral stairwell as viewed by someone riding down the bannister.

Time-lapse photographic drawings   Have the children envision scenes like the hatching of a chick. In a series of time-lapse sketches, have the students depict the process of birth over time.

Symmetry and asymmetry   A discussion of equality, gravity, balance of nature, and mirror images can lead to art activities that differentiate between symmetrical and asymmetrical balance. These activities, in turn, can easily lead to crystallography activities.

Pressure   Have a student do a one-foot stand. Have another student trace the outline of the weight-bearing foot. What pressure per square unit is being exerted on the floor by the student? What is the best way of finding out?

Energy   The human body is an energy system. How many ways might the movements or accomplishments of this energy system be utilized? How does one portray energy? How would one draw a picture depicting kinetic energy and potential energy? What do we mean when we say a body is at rest? How would a child represent this, using lines and form?

Force   What kinds of lines and shapes can students use to create a feeling of force? Graphically show the force of the wind, the ocean, a lava flow, an earthquake, and a paralyzing punch.

Tension   Tension may be represented with constructions using threads, string, or fine wire pulled taut throughout sculptures. Have the children make models using wires or strings to show tension. What happens to the balance of the construct when additional tension is added to any single strand of wire or thread? A discussion of distortion, rupturing, and fracture may quickly ensue. Children will do an outstanding job of mood pictures. Ask them to express tension in a painting or line drawing.

Weight, mass, and thrust   Try to portray mass in motion. A 280 pound fullback crashing through an opposing defensive line conveys the idea of mass in motion. Have the children graphically record this action. Sharp jagged lines can be suggestive of rising, charging linemen. These could be contrasted with the fluid motion and at other times erratic evasive maneuvers of the ball carrier.

Collages with natural materials.   Have the children collect various plants (living or dead), bark, seashells, rocks, minerals, seeds, and anything else. This activity involves observation and possibly classification, depending on the direction in which you wish to move the lesson. Encourage overlapping and balancing the size and texture of the materials used so as to construct collages.

Use discretion with plants. Some plants are poisonous. Some children are allergic to various plants.

Snowflake Fold a square of paper so that a six-sided figure (with three axes of symmetry) can be cut from it. This activity can be used to reinforce the concept of symmetry. Extend it to investigations of crystals. Ice formations on a windowpane can serve as an excellent beginning for this activity.

Flowers and weed shapes Many children can be motivated with flowers or interesting weeds. Use these as a motif for creating patterns for fabrics. Abstract the original shapes; use colors imaginatively. Radial, axial, and nonsymmetrical patterns are extremely interesting to observe and to make. Dehydrating flowers can be an interesting and profitable venture. The construction of paper flowers involves measuring, the concept of surface area, composition, geometric shapes, and color.

Trees, stumps, bushes, and brushes Observe trees. Trees have the general configuration of an umbrella or a cone. Some trees rise upward; some droop earthward. Some are stiff and unyielding; others bend and sway with every breeze. Deciduous trees differ from evergreens. Fruits, blossoms, leaves, and nuts of trees can all be described using the five senses. All of these can be used in classification activities. Leaf prints, made by spatter-painting over leaves (or rubbing them on carbon paper and then placing the leaves on white paper), can be turned into an art project by having students attempt to create unusual patterns by varying and overlapping the prints.

Spider webs Spiders weave webs with such mathematical exactitude that one species is actually called the geometric spider! Honeycombs, beehives, nests, and tunnels are all examples of the clever way in which insects and animals use geometry and artistry in their struggle for existence.

Nature's geometric and artistic skills are everywhere. They are seen in everything around us — the stars, the moon, the earth, and in all that goes to make up the universe. Can you build an awareness of these skills in your creative-sciencing activities?

SCIENCE, SOCIAL SCIENCE,
AND SOCIAL STUDIES

The term *social science* usually includes such subjects as sociology, anthropology, history, political science, social psychology, and economics. The term connotes a conglomeration of disciplines, all referring to various kinds of knowledge concerned with the social relations of human beings. *Social studies*, as differentiated from *social science*, has a slightly different meaning, for it is essentially a curriculum category that often includes the subject of geography. In both subject areas, the concern is with human beings and their social relations.

The area of social studies has much to contribute to the study of science. Geography is the study of place, or space, in the same sense that history is the study of time. Geography revolves around two essential questions. First, where are things located? And second, why are they located where they are? Answering these questions involves mathematics, science, language arts, reading, art, and music. The natural sciences and the social sciences cannot be divorced from one another. The study of human activities, coupled with that of climate, vegetation, soils, and landforms, makes social studies a truly integrated topic.

Distinct areas of social studies that complement science include study of the earth, mapping the globe, the life-supporting layers of the earth, variations in the surface of the earth, the changing earth, people on the earth, and the earth and its spatial interfaces. Each area is rooted in science.

The globe   Try studying the position of the earth in relation to the sun, the solar system, the galaxy, the universe; the motion of the earth, such as rotation and revolution; the shape, volume, and density of the earth; the speed of rotation and revolution of the earth; the inclination of the earth's axis with the resultant seasons and distinct geographic regions; the history of human exploration of the world.

Mapping   The orientation of oneself on the globe involves science and mathematics. The implementation of a grid system, with base meridian and equator lines and the establishment of lines of longitude and latitude, is of prime importance in mapping. Aspects of declination, magnetic inclination, and compass directions are also science topics. Can you design a set of mapping activities for children?

The life-supporting layers of the earth People live at the interface of land, sea, and air. The three parts of the earth — the atmosphere, the lithosphere, and the hydrosphere — are inextricably related. The existence of humanity is influenced by all three parts. Numerous science topics are presented describing these layers. Social studies and geography emphasize people in relation to these three layers.

The life-sustaining layers are playing a more prominent part in basic problems of people's survival. Soils and their relation to food production plus conservation of our natural resources is of critical importance. Consider also weather and climate, the water cycle, the life cycle — even earth and space navigation.

Variation in the surface of the earth This large area of geography deals with the structure of the earth, and it reveals the earth changing under the influence of weathering and erosion.

Resources of the earth The earth's resources are of two varieties — renewable (forest products) and nonrenewable (oil and gas). The latter includes the rocks and minerals of the land and the oceans.

People on the earth The concern of this area is people and their interrelation with the land. People change the land and the land changes the people. The problems of ecology, population, pollution, transportation, marketing, and every human activity, including politics, are involved. People choose where they live, how long they live there, what they do, and for how long they do it.

SCIENCE AND HEALTH

Learn about the human body. Observe changes in size, variations in shape, ability to move, strength to push or pull, ways to communicate, pain to endure, and tastes to enjoy. Learning about the human body — how it is constructed, what it can do, and how it changes — is essential to learning how to care for it. Health and science go together, as these examples illustrate.

The five senses All of science begins with observation. Sight is the sense we rely on most. Introduction to and explanation of the techniques for describing something through the use of the five senses is a necessary requisite for all of learning. The senses serve many other purposes, from giving enjoyment to being prerequisite for survival.

The circulatory system   Have children investigate their hearts and how they beat. Several questions for investigation might include:

— What is each student's number of heartbeats per minute?
— How do the data on the number of heartbeats per student compare?
— What might account for variations in students' heartbeats? Is weight a factor? Age? Sex? Or an interaction of all these factors?
— Is the pulse rate the same as the rate of heartbeat?
— What activities would increase the heart rate?
— What activities would decrease the heart rate?

The skeletal system   Have the children construct a chicken or squirrel skeleton. Have them compare the functions of their bones to those of the chicken or squirrel. Do human beings have more bones than chickens? How does the skeletal system aid other systems of the body?

The muscular system   Place your arm on a table top and place a soft rubber ball in the open palm of your hand. Then, repeatedly squeeze the ball. Record the number of times per minute you squeeze the ball. Continue recording until you are stopped by muscle fatigue. How did your hand feel when you started? How did your hand feel when you stopped? Did the number of squeezes per minute decrease with time? Can you prepare a graph of your data?

Compare the data you collected with results obtained by children. What might account for the differences? Compare the data on an adult to those on a child. Do you predict any differences? What variables could be manipulated? Could you predict what would happen on another try after a few minutes' rest?

Many other aspects of health such as body motion, protection and survival of the human race, growing and changing, alcohol, tobacco, drugs, and the voice box and windpipe (that control pitch and volume) are all excellent examples of topics that can be used in skill activities that relate science and health.

Think of three creative-sciencing skill activities that could be undertaken with children, using the topic of the nervous system.

SCIENCE AND READING
AND THE LANGUAGE ARTS

Creative-sciencing teachers recognize that instruction in reading and language arts usually accounts for up to half of the instructional time in schools. They capitalize on this by including motivational science activities in the reading and language arts program. Section 3 includes many examples of skills and techniques that work well with the reading and language arts program — task cards, puzzler activities, audiotutorial tapes, invitations to investigate, skillettes, science-book kits, games, modules, and interest centers. Studies show that children prefer to read books that include science topics.[2] It behooves teachers to capitalize on this by allowing children to select their own reading materials.

Science is a natural to develop the following aspects of reading and the language arts:

Vocabulary The technical vocabulary of science is replete with terms that have Latin and Greek roots and affixes. Knowledge of these word parts often assists children in understanding and retaining this vocabulary. Prefixes such as *bio-*, *micro-*, *radio-*, and *aqua-* can go with roots such as *-meter*, *-nuclear*, and *-flex*. Have children combine prefixes, roots, and suffixes to make words, look up their meanings, and relate them to the real world by using the words in a sentence or two.

Cause-effect Science deals with many cause-and-effect relationships. Prepare lists of diseases such as *measles, rabies, tetanus, cold*, and *ringworm*, and have the children decide if each disease is caused by a virus, bacteria, fungus, or protozoan.

Following directions This crucial skill is needed in all subject areas, especially science, since science texts contain so many directions for completing experiments and investigations. To assist children in learning to follow directions, select a science experiment and place each step in the directions on a note card. Order the cards by writing a number on the back of each one. Mix up the cards and have the children figure out

2. Gerald H. Krockover and Kay Brown, "A Reading Preference Test: Rationale, Development, and Implementation," *Elementary English*, vol. 51, no. 7 (1974), pp. 1003–1004.

the order of the steps needed for the experiment. Have them turn the cards over to check the sequence.

Factual descriptions Science textbooks contain many factual descriptions of physical phenomena. One way to encourage children to become familiar with factual descriptions is to select short passages from science books and place them on index cards. A child selects one of the cards, reads it, and puts it back. Then, the child is asked to state the subject of the passage, to relate two or three facts from the passage, and to summarize the passage.

Fantasy trips Science is a natural for stimulating children's imaginations. Create a space journey for your class by dividing the class into groups of three or less and giving each child a role such as the space doctor, the astronaut, and so on. Have children act out the roles and pretend that they are taking the space trip. Darkening the room and playing space music can make the trip more realistic.

Make a calculator spell Have children use their calculators to solve a problem and then "read" the answer. Ask the question: What animal lays golden eggs? The children then add, subtract, multiply, and divide to arrive at the answer *goose*! On the calculator, certain numerals when inverted resemble letters. For example, 1 stands for *i*, 3 stands for *e*, 6 for *g*, 8 for *b*, 0 for *o*, 4 for *h*, 5 for *s*, and 7 for *l*. With these numbers it is possible to form a variety of words.
    Question: What is on the beach?

1. Start with 10,000 grains of sand.
2. Add 77 crabs.
3. Subtract 70 crabs because a fisherman caught them.
4. Subtract seven more crabs because the fish ate them.
5. Multiply by seven sunbathing people.
6. Add 7000 more grains of sand.
7. Add 345 frankfurters to eat on a picnic and then turn the calculator upside down.

The answer is *shell*.

Sequencing Science television shows can help children learn the skill of sequencing. Tape record a short segment from a TV science show to set the scene. Place each line of the script on a strip of paper, mix the strips up, and then have children order the scene into the proper sequence. Also give practice with predicting by having the children write the script for the next episode.

Using textbooks and library books

When using a science textbook in the elementary classroom, the teacher must recognize that the reading levels of the children vary greatly. As a rule of thumb, the number of levels will equal the number of the grade level plus or minus 1. That means a third-grade teacher can expect up to four reading levels in the classroom. As the grade levels increase, so does the range of reading levels. As a result, a single science textbook cannot serve the entire class. Creative-sciencing teachers use the practical approaches suggested here to meet the needs of their students rather than having all students read the same material in the same manner. The average science textbook contains reading levels from grades 2 through 12. It is recommended that teachers always read the textbook first and then decide which students ought to read which material. Students need to be taught how to use reference and resource materials, interpret context clues, recognize symbols, identify pertinent details, select main ideas, and draw conclusions.

Conceptual themes can also serve as a basis for relating science to reading-language arts. Children will enrich their science experiences by selecting library books that relate to such science themes as change, diversity, relationships to one's surroundings, and so on. Books related to these themes can be selected for a variety of interests and reading levels.

Children should be encouraged to use the science material they have learned by writing their own fiction. Combining science background with creative writing serves to reinforce the science content and to develop critical and creative thinking abilities. *Charlotte's Web* by E. B. White (Harper and Row, 1952) can serve as an excellent beginning for a discussion about writing books with animals as the main characters. Other science themes and related books that can be used to motivate creative writing experiences include:

— Machines: *Chitty Chitty Bang Bang: The Magical Car* by I. Fleming (Paragon, 1964)
— Chemistry: *The Sorcerer's Apprentice* by P. Dukas (Random House, 1974)

A POEM ON THE SKIN
by Dave

Once your a tootsie,
You jump on a footsie
Just follow your skin,
And you'll come to your shin
When you get on your knee,
You act like a bee
When you come to your thigh,
Just sigh.
When you get to your waist,
You eat up some paste
Then you jump on your belly,
And eat some jelly
Then you walk to your arms,
And put on some charms
When you come to your head,
Go to bed.

SOME SKIN
by Kay

Some skin has fur . . .
Some has hair . . .
Some skin is dark . . .
Some is fair . . .
Some heads are bushy . . .
Others are bare . . .
Fish skin is scaly . . .
I do declare . . .
But as long as I have it,
I don't care.

— Weather: *Miss Pickerell and the Weather Satellite* by E. MacGregor (McGraw-Hill, 1971)
— Space: *The Three-Seated Universe* by L. Slobodkin (Macmillan, 1974)
— Occupations: *Uncle Bill's Ice Cream Shop* by G. H. and S. D. Krockover (Vantage Press, 1978)

Once motivated, the children may also enjoy writing poetry to complement science activities they have just completed. The two poems on page 149 were written by fifth-grade children and have not been edited.

The "Brainstorming in Science" (bis) section of *Creative Sciencing: Ideas and Activities for Teachers and Children* contains 129 science activities that relate to other areas of the curriculum. These can assist you in developing an integrated curriculum.

PAUSE FOR A SUMMARY

— Successful science instruction at the elementary level can be accomplished by teaching science as a discrete topic and/or by teaching science as it weaves through the entire elementary curriculum — mathematics, art, social science, social studies, health, and reading and language arts.
— Science as promoted through inquiry necessitates the collection of data.
— Distinct areas of social studies that complement science include the study of the earth; mapping the globe; recognizing the life-supporting layers of the earth; identifying variations in the surface of the earth; discovering concepts about the changing earth, people on the earth, and the earth and its spatial interfaces.
— Creative-sciencing teachers use science activities to develop these reading and language arts topics: vocabulary, cause-effect, following directions, factual descriptions, fantasy trips, calculator communication, and sequencing.

What they neglected to tell me about...

Does it matter?

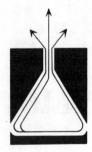

What you don't know will hurt you.
— De Kro

THE FIRST DAY

Even for senior staff members, the first day of school is always an adventure. New books, new supplies, and new students all add to the excitement. All teachers privately ponder the arrival of those students who become "theirs" for a period of one year.

Teachers usually are concerned about the class they inherit; the children, in turn, are concerned about the teacher they inherit. It works both ways. Yet in your favor is the fact that regardless of your personality, intelligence, and so on, the children you are assigned accept and defend you loyally to the end. Good or bad, you belong to them (at least for a year). And, no matter what, they desperately want to please you.

Some cautions for the first day

Allow yourself appropriate lead time long before the first day arrives. Plan the first three weeks well in advance. The more prepared you are, the more relaxed and resilient you will be. In teaching, there are few substitutes for good organization and planning. Plan sufficient science instruction in detail. This instruction should be supported by appropriate activities, some of which extend over the full term of your science unit. Suppose the science topic is weather and you wish to develop these concepts: Air expands when heated. Air has mass. Air exerts pressure. Your instruction of weather and these concepts might be supported by such activities as the construction of weather instruments and a weather station. The weather station might contain a weather vane, a wind speed indicator, a deflection anemometer, a pressure tube anemometer, a rain gauge, and other equipment that will involve students in activities supporting your instruction in weather concepts.

Be prepared to involve the children the first day in some exciting activity. Don't spend the entire day with so-called housekeeping chores such as passing out books, paper, or instructional notes. Inasmuch as the first day can be your best shot at setting the tone for the entire year, pick a winner of a science activity and kick it off with gusto. Don't make it too hard, too involved, or too long. A short, high-interest activity to set the stage for an action-filled year will do the trick. If you start with a unit on weather, you might well introduce the unit by having a kite or paper airplane flying contest.

Start the year off by presenting a firm image of yourself. You can always soften up as the year winds down. If you start loose and try later to tighten up a bit, you may find it difficult to recapture lost ground.

Establish a few essential ground rules, but don't overdo it. Consider these examples: Everyone is responsible for his or her own behavior. Science equipment is to be used correctly, then cleaned and stored in its proper place.

Don't allow yourself to get into arguments with the children over such impatient reactions as, "Do we have to repeat the experiment over and over?" Provide the students with your rationale, explaining the necessity for what is to be done and then ask them to do it. You are in charge. You know it. They know it. And they know you know it.

Be confident. Don't worry about making mistakes; everyone makes a good share of them. Not every experiment works exactly as it is supposed to; not every mothball placed in baking soda and vinegar rises and sinks; and so it goes. Be flexible. In teaching, the unexpected should be the expected. A seemingly simple assignment, such as planting seeds in a tumbler garden, can have a wide range of results — from the growth of a healthy plant to the development of a vigorous mold. Unexpected results raise the question "How come?" and sometimes generate more interest than anticipated results. Make use of them. Resiliency is your buffer for maintaining your equilibrium. You must be like a reed in the wind, bending and rebounding.

Don't be overly concerned with criticism. No other profession is as susceptible to criticism as teaching. Listen to criticism. Discern what may be constructive in it. Cast aside the trivial and work on the constructive with this goal in mind: to become a better teacher every day in every way.

Take some calculated risks. Try those teaching innovations that you have been wondering about. (Do they really work?) Grouping, independent work, open-concept learning, the use of numerous divergent questions, open-ended experiments in science, the use of science-theme interest centers are all exciting possibilities for improving and enlivening science instruction. Try some of them. Remember — little risk, little progress.

Don't waste time. The most precious thing students bring to your class is time. It should not be squandered. If you waste five minutes an hour in a six-hour day, five days a week, 180 school days a year, you waste 75 hours a year or about twelve six-hour teaching days per individual per year. Work at staying healthy. Remember, you are probably the oldest (as well as the dearest) thing in the classroom. Protect your health from the continuing onslaught of typical classroom illnesses by getting lots of rest. Enthusiasm and vitality are two of the most important ingredients in good teaching. Nourish them continuously and they will flourish.

Creative-sciencing teachers:

- Give positive recognition.
- Assure each student some success.
- Involve students in making choices and setting rules for acceptable behavior.
- Involve students in helping each other.
- Plan with students.
- Make the classroom a special place.
- Encourage student-constructed materials.
- Encourage independent problem solving.
- Provide activities to help each child take pride in his or her contributions.
- Develop freedom with responsibility.
- Involve parents, principals, and anyone else who will help.
- Assign homework with choices.
- Emphasize mastery of skills.

— Spice teaching with humor.
— Show enthusiasm.
— Work hard!

DEMONSTRATIONS TO EXPERIMENTATIONS

A classroom demonstration is a controlled performance, usually given by the teacher, to present some preselected phenomena. The outcome, barring unexpected results, is generally known. Valid arguments in favor of a well-planned and well-executed science demonstration include: expediency of time, broad coverage of content, instruction in safety in science, broadcasting of procedures, and in general setting the stage for future instruction. Caution, however, must be exercised to avoid overuse of demonstrations in lieu of a more student-centered approach.

The egg in the bottle, an air pressure activity, is an example of a routine science demonstration. To present it, you need a shelled, hard-boiled egg, a glass milk bottle, a sheet of newspaper, a match, and water. The newspaper is rolled into a wad so that it can be ignited and dropped to the bottom of the milk bottle. The burning paper will eventually go out. The heated air inside the glass container will expand, become lighter, and escape, making the air pressure inside the bottle less than the pressure outside the bottle. Moisten the hard-boiled egg. Quickly seal the mouth of the bottle by positioning the wet egg, blunt end first, in the mouth. This will initiate the egg's motion. If the egg is not wet, the rim of the bottle should be moistened before positioning the egg. Water serves as a lubricant and the egg should slip down inside the bottle — sometimes slowly and sometimes with a resounding thud.

The success of this demonstration depends on finding an egg the approximate size of the milk bottle opening. The cost of eggs as well as the difficulty in obtaining 10–15 glass milk bottles (the kind that Grandma used works best) with the right-sized mouths make this activity better for a teacher demonstration than a student activity.

This demonstration is a good introduction to a study of air pressure. The very act of getting the egg intact inside the bottle is a bit of a mind boggler. Few people think it can be done. Accomplishing a seemingly impossible task is interpreted as a discrepancy to a viewer; it is contrary to expectation. These discrepancies are sometimes called *discrepant events*. Getting the egg out of the bottle, intact, further enhances the discrepancy. This can be accomplished by carefully adding sufficient water to the bottle to thoroughly rinse out all the burned and unburned debris. Also, the water acts as a lubricant for the egg's exit. After all the water and debris are removed, turn the bottle upside down and position the egg so that it rests in the neck of the bottle. Place the bottle directly over your head. Using your thumb and first finger, form a circle against the mouth of the bottle. Now, place your mouth against your circled fingers, and blow into the bottle as hard as you can. Be careful! Step lively! The egg can exit slowly, or it can come out rapidly and land on you. In either event, the egg in the bottle is a dramatic demonstration that can kick off your unit on air pressure.

Demonstrations have their place. When necessary, use them. Remember, however, the payoff for children is in their personal mastery of the skills of experimentation. Every demonstration you perform makes you, not the students, more adept. While a demonstration may occasionally be necessary, it is not sufficient for student learning.

True scientific experimentation in the elementary school is difficult. It rarely happens, but that does not mean it couldn't. Nor does it mean that it shouldn't. It just rarely happens. Experimentation calls for knowledge about the problem under consideration, as well as perseverance and fastidiousness, all housed in an open and creative mind. This is a lot to expect of elementary pupils. The best we can do is to involve them in a variety of science activities that approximate experimentation. These activities must correlate with the grade level and background of the children. Yet, all children need initial and sustained exposure to the experimental process (in part or in toto) to acquire a familiarity with the concept of experimentation and the skills associated with it. The concept of experimentation is complex. Its acquisition for most children will span several, or all, of the elementary school years, provided of course that continued instruction in the process is given.

An experiment usually seeks an answer to a question through an investigation of some observable phenomena. Experimentation involves:

— the delineation of a problem;
— the construction of hypotheses;
— the identification of pertinent variables;
— the organization of a design compatible with the formulated hypotheses;
— the controlling of variables;
— the act of collecting data;

— interpreting the data;

— summarizing the conclusions in light of the problem.

All of this is somewhat removed from either a teacher demonstration or a simple, teacher-student involvement in a specific science activity.

Children's abilities to "experiment" may be improved through the use of either teacher-structured science activities or unstructured explorations. In either case, a problem needs to be identified. Children usually need assistance in identifying and stating problems. You, as the instructor, can help them phrase problems so that they are manageable, understood by all, and experimentally explorable.

The planting of seeds in the "Tumbler Garden Has It All" activity is a good example of an unstructured approach. Have all the students plant tumbler gardens that are equal in the following ways: size of containers; size, variety, number, and position of seeds within the containers; size of paper toweling; folding of paper toweling; positioning of paper toweling within the containers; quantity of wadding; and amounts of water.

In the process of establishing equality among the tumbler gardens, you could have the children determine the volume of the containers, observe the seeds, describe the seeds externally and internally, make inferences about the seeds, predict some behavior of the seeds, classify the seeds, determine the mass and the hardness of the seeds, measure and record the amount of water furnished the seeds, and so on. All of this is valid sciencing and accumulates knowledge of seeds, but it is not an experiment. An experiment takes place in the pursuit of an answer to a question:

— Does light make a difference in the growth of seeds?

— Does temperature make a difference in the growth of the seeds?

— Does the amount of water added affect the growth of the seeds?

In establishing the question you wish to pursue through experimentation, avoid *Why* questions. While such questions spark interest, they usually encompass too much. Why are plants green? is an excellent question but a formidable one for elementary school children. *Who, What, When, Where,* and *Does* questions are usually more manageable than *Why* questions.

Once the question has been posed, pertinent variables (which when modified will produce some change in the system) must be identified.

Identifying them is a critical point in determining the success of the experiment. Too often in elementary science, the variables are not controlled. In the tumbler garden activity, suppose the seeds were planted with a multitude of variables operating. Not only would the activity fail to answer the question originally asked, it would conclude little and undoubtedly raise more questions. If this is your purpose — and on occasion it may be — fine! And if, at some point in the developmental sequence of the experiment, it is your intent to limit the activity, that would be all right too. But if you are going to engage in an experiment to answer a specific question, you must control all variables except the one you are investigating, the *manipulated variable* (independent variable). Variables that respond to this manipulation are called *responding variables* (dependent variables). Once you have controlled the variables, the act of undertaking the experiment — collecting and interpreting data, and so on — more readily falls into place within the scheme of the total experimental process.

Why a *why* question sometimes fails

X, WHY, AND THEE

The day lay heavily that Friday afternoon. The noonday lunches, resting comfortably like small barbells in the stomachs of 30 sixth-grade science students, were reflected in the eyes of the students. Undaunted, the elementary science teacher commenced teaching science.

"O.K., watch me carefully," cautioned the instructor. Out of a mysterious container came a smoking substance identified by the instructor as dry ice. "Is everyone watching?" queried the instructor. Injecting some mid-day humor, he added, "Notice that during the entire demonstration no finger ever leaves the hand." This was followed by the striking of a match. It was placed in close proximity to a piece of dry ice. It went out immediately.

"Can anyone tell me *why* this happened?" asked the instructor. He repeated, "Why do you think this happened?"

Adapted from Alfred DeVito, "X, Why, and Thee," *Science and Children,* vol. 8 (November 1970), pp. 24–25. By permission from the National Science Teachers Association. Copyright 1970.

Almost everyone in the classroom saw what had transpired. The match went out. The "Why" was something else.

The teacher sensed the partial vacuum buildup. It moved out from the back of the classroom like a tidal wave. Buttressing himself against the cresting wave, he slowly sank, gurgling, "Why? Why? Why?"

If one could turn on a yet to be invented Instantometer that could flash the students' thought in response to *why*, the responses would be varied. The responses might range from, "It went out because it was damp." "It went out because it blew out." "It went out because it simply burned out." And, "I don't have the foggiest notion of *why* it went out." The last response would probably be the most common response.

We do not believe any of the responses would have satisfied the instructor. The instructor was probably pointing to one scientific explanation for *why* the match went out.

How many times have you asked the students in your class "Why?" How many times have you read in the eyes of your students the sign, "Closed for renovations"? How does one move students from that "closed for renovations" look to an "open for innovations" look?

The instructor in this case made use of a good discrepant event. Inquiry was inherent to the demonstration. What might have gone wrong? Did the *why* come too soon? Perhaps the jump from observation (the eye) to explanation (the why) was too great for the students.

A problem-solving approach wherein the instructor leads the students to the answer to a *why* question through analysis and synthesis is preferable to performing a demonstration and simply asking, "Why?" The students should be trained to "stop, look, and think" before answering *why* questions. They should be trained to analyze the components of the activity. They should be able to ask themselves the following questions: What is involved in the activity? What characteristics do the objects involved in the activity have? How do the objects involved in the activity interact with one another? What might one hypothesize about these objects observing the results of the demonstrations? How can one set up an experiment to check these proposed hypotheses? How does all this information aid in answering the question, Why?

This approach assumes that the students have had some training in working towards the solution of problems. If they have not had this training, a discrepant event presented in isolation may draw a blank from the audience. If it does, be prepared with supportive sequential demonstrations that lead students to knowledge which may allow them to understand the *why* of the discrepant event. Consider these demonstrations:

1. Expose a piece of dry ice to room temperature.
   Observation: Under ordinary conditions, the dry ice changes directly from a solid to a gas without passing through a noticeable liquid state.
   Question: What happened to the dry ice? Where did it go?
2. Taking the necessary precautions, put dry ice in a metal pitcher. Let the dry ice evaporate. Pour the contents of the pitcher into a paper bag filled with air. Attach it to one side of an equal-arm balance (a balanced yardstick or dowel will do), and balance it with an equal size bag also filled with air.
   Observation: The paper bag to which the generated gas has been added moves downward.
   Question: What does this reveal about the generated gas?
3. Again evaporate some dry ice in a metal pitcher. Construct a cardboard trough. Incline the trough towards a burning candle or a series of burning candles. Pour the contents of the pitcher into the trough.
   Observation: The generated gas flows downward and extinguishes the candle.
   Question: What does this reveal about the generated gas? Is it lighter or heavier than air?
4. Wrap a thin wire around a lighted candle and lower the candle into a metal pitcher filled with air. Repeat this, lowering the burning candle into a metal pitcher in which dry ice has been evaporated. Compare the two activities.
   Observation: In the pitcher filled with air, the candle continues to burn. In the pitcher filled with the generated gas, the candle quickly goes out.
   Question: Will this generated gas support combustion?

The questions in demonstrations 1–4 build a sequence that leads to the explanation of *why* the burning match went out when placed in proximity to the dry ice.

Why questions can be difficult. The biggest fear of many prospective teachers seems to be that students will ask them questions like these:

Why is the sky blue? Why can't animals talk? Why do snakes shed their skins?

Students need time to reflect on the *why* of things. Students need prerequisite knowledge to make an instantaneous response. If this is lacking, students need a strategy and time to employ the strategy to arrive at a response to a *why* question. If students lack a strategy, concrete, sequential demonstrations or activities should be provided which lead them to a solution of *why* problems. Uncover the pieces and permit the students the opportunity to weld ideas together. Try it. WHY NOT!?

points to ponder

Select a creative-sciencing activity and present it as a teacher demonstration. Next, modify the same activity and approach it as a creative-sciencing experiment.

HOW TO ORDER AND OBTAIN MATERIALS

While elementary school teachers of science substitute common household items such as mayonnaise jars, aluminum pie plates, medicine bottles, and eyedroppers for commercially produced science equipment, some items like magnets, timers, magnifying glasses, glass tubing, and stoppers must be purchased periodically.

Before ordering science materials, compare quality and prices in the catalogs of different general science supply and equipment organizations. Before requesting catalogs and prices from various distributors, check with middle, junior high, or high school science teachers in the school district. Usually, one or more of these individuals will have catalogs and price lists available, thus saving you time. If not, the following list may help.

Supply houses for science supplies and equipment

— Carolina Biological Supply Co., Burlington, NC 27215
— Central Scientific Co., 2600 South Kostner Ave., Chicago, IL 60623
— Connecticut Valley Biological Supply Co., Inc., Valley Road, Southampton, MA 01073
— W. H. Curtin Co., Box 1546, 4220 Jefferson Ave., Houston, TX 77023

— Fisher Scientific Co., 1458 North Lamon Ave., Chicago, IL 60651
— General Biological, Inc., 8200 South Hoyne Ave., Chicago, IL 60620
— Hubbard Scientific Co., PO Box 105, Northbrook, IL 60062
— LaPine Scientific Co., 6001 South Knox Ave., Chicago, IL 60629
— Macalaster Scientific Co., Rt 111 and Everett Turnpike, Nashua, NH 03060
— Nasco, Fort Atkinson, WI 53538
— Sargent-Welch Scientific Co., 7300 North Linder Ave., Skokie, IL 60076
— Southern Biological Supply Co., McKenzie, TN 38201
— Southwestern Biological Supply Co., PO Box 4084, Station A, Dallas, TX 75208
— Trans-Mississippi Biological Supply, 892 West County Road, B, St. Paul, MN 55113
— Ward's Natural Science Establishment, Inc., PO Box 1712, Rochester, NY 14603; *or* PO Box 1749, Monterey, CA 93940

Suppliers of unusual plants and unique seeds[1]

— A. E. Allgrove, North Wilmington, MA 01887 (specialists in terrarium plants, ferns, mosses, and so forth)
— Buell's Greenhouses, Eastford, CT 06242 (specialists in Gloxinias and African violets)
— Burnett Brothers, Inc., 92 Chambers St., New York, NY 10007
— W. Atlee Burpee Co., Philadelphia, PA 19132 (seeds and equipment)
— Carolina Biological Supply Co., Burlington, NC 27215
— Farmer Seed and Nursery Co., Faribault, MN 55021
— The House Plant Corner, Box 810, Oxford, MD 21654
— Merry Gardens, Camden, ME 04843
— George W. Park Seed Co., Inc., Greenwood, SC 29646

Microorganism suppliers

— Carolina Biological Supply Co., Burlington, NC 27215; *or* Gladstone, OR 97027. (viruses, bacteria, fungi, algae, and protozoa)

1. Common plants and seeds can be found at local greenhouses.

— General Biological, Inc., 8200 South Hoyne Ave., Chicago, IL 60620. (bacteria, fungi, algae, and protozoa)

points to ponder

Find out where in your area you can obtain science materials needed for your classroom and prepare your own materials list. Try businesses, industries, other schools, colleges, and universities. Prepare a list of common household items that you could recycle and use in your creative-sciencing program.

ACQUIRING, ORGANIZING, STORING, AND DISTRIBUTING SCIENCE EQUIPMENT AND MATERIALS

What to get and where to get it

The "Shoestring Sciencing" section in *Creative Sciencing: Ideas and Activities for Teachers and Children* provides you with many ideas for obtaining the equipment (or suitable substitutes) as well as the materials necessary to successfully transform "reading about" science to "doing" science.

Some basic pieces of equipment are an absolute necessity: balances, timers, some means of magnification (lenses, microscopes), measuring devices (rulers, thermometers, graduated cylinders), Pyrex glassware (test tubes, flasks, beakers), and heat sources (hot plates, propane tanks). Although some of these basic pieces, such as balances and measuring devices, may be fabricated, most essential pieces must be purchased.

Once you have acquired the basic equipment, you can begin to assemble sundry materials including vials, eyedroppers, weights, geometric shapes, containers, wheels, gears, mirrors, dowels, straws, corks, wire, string, screening, seeds, glass rods, marbles, and balloons. (See ss 2, "Recycling Helps Science Education," in *Creative Sciencing: Ideas and Activities for Teachers and Children* for the countless other items you may need.) All these materials are readily available in local hardware stores, grocery stores, department stores, toy counters, friends' attics and basements, salvage yards, and many other places. Acquiring the wide variety of materials that makes a science program "fly" is the first step in providing real hands-on sciencing for children.

Getting organized

The next step, organizing this equipment and material, is crucial. All that you have acquired is useless if it is not organized so that it is readily available. Organization expedites your science instruction. Time is the most precious commodity that children have. To waste it would be improper. The best-conceived science lesson loses its effectiveness if it is confounded by delays and shortages of equipment. Discipline problems fester in the time between instruction and direct involvement. The longer this time, the greater the inducement for children to engage in activities other than those you have planned. Organization is prime. It is part and parcel of good planning.

Places to store materials

You will need space for storage. Some old guru once said, "Every graduating elementary education major should get a sheepskin, keys to a warehouse, and a strong burro." While it may at times seem as though you need a warehouse (and occasionally a burro), lesser space will do. Using your investigative talents, ferret out all the nooks and crannies in your school. A closet or an old cloakroom will do. If your school is not so endowed, a corner of your classroom will suffice. Shelves would be ideal, but egg crates, orange crates, or bricks and boards can be substituted for them. The only constraint on finding space is your desire to achieve the goal. Don't weaken. The end result is well worth the trouble.

Having discovered or created a storage area, you will need to make some decisions as to what equipment and material falls into the categories *valuable/not valuable* and *storable/not storable*. Valuable and dangerous items should be under lock and key. You'll need a foolproof backup system for an extra key. (Many a fine science lesson has gone awry for the want of a key.) Large pieces of equipment — balances, for example — should be out in the classroom. Balances, microscopes, and timers should be common, visible pieces of equipment. Children should be encouraged to use them as frequently as they use the pencil sharpener.

Ways to distribute materials

Small science materials can be organized in alphabetized boxes or trays — balloons would be in the B box or tray, and so forth. This method works well for some teachers who compile from their science lesson plan a materials list and, prior to the instruction, assemble needed materials in trays ready for distribution to the students. Other teachers prefer to organize lessons in boxes. Sometimes this is referred to as shoe box sciencing. A shoe box labeled "magnets," for example, would contain magnets, wire, nails, thumbtacks, paper clips, steel filings, and anything else needed for a lesson on magnets. All the teacher needs to do is pull out the appropriate shoe box and what is needed to complete the lesson would be ready. This method is expedient, but it is somewhat inflexible inasmuch as all lessons are predetermined and deviations are usually limited. Nevertheless, shoe box sciencing does work.

Either of the two methods breaks down when expendable items are used up. Thus, each shoe box or tray needs to have a record card that lists the expendable items and the quantity that should be included in each container for a lesson. An inventory of the containers before the lesson will determine the current status of the expendable items.

It is advantageous if each student can have a full set of items to accomplish the objectives of the lesson. If this is not possible, keep the number of students using one set of equipment as small as possible. Also, when possible, have the children personalize the items they use. Have them print their initials or names on their balances, their baggie gardens, their baby food jars. Elementary school children respond favorably to this identification. Some teachers have each child keep a personal science cigar box with often-used items like tweezers, pins, rulers, hand lenses, and eyedroppers. This saves on the time spent distributing these materials. Also, children treat these items with greater care.

You can do it — with a little help

Having read all this information about the acquisition, organization, storage and distribution of materials and equipment, you may say, "See? That's why I am reluctant to teach science!" This process does not have to be tedious. Again, the key is your ability to organize. Organize parents, organize the children and, if possible, organize the principal. They can all help you procure, clean, keep track of, and sustain your organization of materials. Children enjoy meaningful tasks that help you and them.

SAFETY IN SCIENCE

*teaching science well is teaching
science safely; no one wants to invite
safety problems; to be forewarned
is to be forearmed!*

Some cautions for teachers

Try all activities and demonstrations beforehand. If you have any doubts about safety, make necessary revisions to eliminate the hazards.

Your school may have some safety rules concerning the use of fire, wearing eye protection equipment, electricity, and so on. Always check with the principal or science supervisor before initiating any science activity that involves the use of an open flame, electricity, and other hazardous materials. When permission is granted, forge ahead, observing the appropriate safety precautions.

Not all hazards can be totally eliminated. In some activities — "Why does a burning match curl up?" for example — burning matches must be used. If hazards exist, provide the students with special instructions, highlighting the potential hazards and the precautions to be observed. This applies to items such as hot plates, propane tanks, soldering irons, electric wires, microprojectors, and audiovisual equipment. Many heated items retain heat long after use and therefore pose special problems in handling, removal, and storage. The heated items should all be allowed to cool thoroughly before moving them. While they cool, however, they need to be identified by signs that warn the children that they are hot. Sometimes a small barricade is necessary to supplement this warning.

Constant supervision in science activities is a must, particularly when using equipment that might present safety problems. This supervision should take into account the number of pieces of equipment, the size of the class, and the length of time groups or individuals need to use the equipment to complete a task. Time constraints, as well as limited equipment, force children to rush. Rushing contributes to accidents. Limit the size of the groups to correspond to the available equipment.

Never suggest that students extend the activity beyond the level to which you take it unless you review and approve the action. Unsupervised experimentation at home or in school can be hazardous.

Everyone feels a bit foolish when an accident occurs. Students sometimes have accidents and suffer in absolute silence to avoid embarrassment. Insist that all accidents, no matter how small, be reported to you. Don't attempt to treat injuries. Refer injured students to the nurse or the principal for further action.

Dull tools are more dangerous than sharp tools. The dull tool forces the user to apply force which can result in an accident. Be sure that all tools are in good condition and that students know how to use them.

Safety with animals and plants

Live animals in the classroom, while delightful, can be hazardous. The nice, gentle animal at home is not the same animal when it is besieged by thirty children vying for its attention and affection. Some precautions are in order.

All mammals should be inoculated for rabies prior to being brought into the classroom. (This requirement alone allows you a way out of agreeing to a child's request to bring in his or her aardvark.) Restrictions should be imposed on bringing to class wild animals such as snapping turtles, rabbits, wounded birds, or poisonous snakes. In fact, discourage bringing pets to class. It isn't fair to the animal — or you.

If, by chance, you weaken and permit small animals in the classroom, observe these simple rules: Provide clean, adequate living quarters free of contamination and secure enough to confine the animal. Plan for heat, light, water, food, and proper waste removal — even on weekends and holidays. Limit the handling of animals. Sanitary measures (washing hands after handling animals) must be practiced. Do not allow the animals to be teased. Any animal bite or scratch should be promptly reported and treated.

Plants, while seemingly harmless, must be viewed with some caution. Don't bring in poisonous plants. Don't allow children to eat or place portions of plants in their mouths or rub them on their bodies. Adverse reactions can occur.

and remember

— All spinning objects are potentially dangerous.
— All items under tension (rubber bands, springs) are potentially hazardous.
— Heating a gas or liquid in a confined container is dangerous.
— Always add acid to water, never the reverse.
— All reagents should be properly labeled and stored in a cool, dry

place (not the refrigerator). Also, combustibles should be stored in a locked metal container.

— Only Pyrex or heat-treated glassware can be safely heated.

— Bacterial cultures can contaminate unless sterilizing techniques are used.

Knowing all this and always remembering to act prudently, enjoy yourself and teach science. We hope that, rather than turning you off to science with all these precautions, we have turned you on to creative-sciencing safety.

points to ponder

Teacher A says, "I don't teach science because of all the safety hazards: I don't want to be sued."

What is your response?

FIELD TRIPPING

Someone once said, "There aren't many wondrous events in a lifetime, but a few will suffice." If properly planned, a field trip can be classified as a wondrous event. Proper planning and attention to some basic concerns can insure success.

Why take them?

A field trip means many things to many people. For some, it may be a quiet walk to view autumn leaves that fringe the school yard. For others, it may be a three day trip to the Royal Gorge. Regardless of the type, extent, or duration, most field trips have many similarities. One important consideration always has to be: Why take field trips? This question has a variety of answers.

It is your turn. Everyone in the school system has visited the ice cream factory, the post office, the zoo, the local bakery, the local newspaper, and the police station. Your students are asking, "When are we going someplace?" But just going someplace is not sufficient. Your field trip should serve a purpose. Perhaps it is to collect some rock-forming minerals to reinforce your current science instruction or to observe dinosaur models at a museum. The major purpose is to give students the kind of firsthand experience not obtainable in the classroom.

You see your students in a new setting. This is usually reflected in a set of behaviors different from those in your classroom. The students see you in a new setting. You look and perhaps act differently when viewed in the field by children.

Some precautions

If possible make a dry run of the field area with a select group of students. Observe their reactions as they peruse the site. Your ability to interpret a wrinkled nose, an impish glint in the eye, or a twitching ear will enable you to forecast your whole class's reactions. Step back yourself and view the area with a critical eye. Think of it under the worst possible conditions. These conditions should dictate the kind of equipment, clothing, and other things that you should have your students bring. Take slides of the area: include land marks, trail signs, places to see, things to do, things not to do, and items to be observed or collected. If possible, bring back actual samples of collectible items. Show the slides to the children prior to the actual visit. These slides and later shots of the group are very useful in the debriefing discussions that follow field trips. Make arrangements well in advance if you are planning to go to a commercial area. Arrive with your confirming letter substantiating the arrangement. Have a name to ask for in case all does not go well.

Common complaints of field trippers are: They can't see. They can't hear. They can't keep up with the traffic. Solutions can be found for these problems. Distributing maps which have labels of the field trip stops as well as printed details of what to see at each stop can eliminate some of the complaints. Dividing the class into smaller groups led by informed adults who can monitor the movement and instruction is also helpful. Including on the map the location of rest rooms, eating facilities (type and price range), and recreation areas will eliminate other complaints.

Even if you can squeeze everyone on one bus, it is a good idea to have at least one parent driving a backup car behind the bus. Many emergencies or "early-return-homers" can be handled more readily using this backup system.

Bring emergency, compact, lightweight, highly nutritious food such as space sticks, chocolate bars, or breakfast squares; a few basic medical supplies such as Band-Aids and insect repellent; cameras; containers to collect things in; plastic garbage bags; tissues; and a can or bottle opener.

Field Tripping / 169

Spell out the rules of conduct. Don't deviate from your prescribed trip behavior. Remember, you are not responsible for minding the lunches, nor are you the keeper of purses, wallets, cameras, or contact lenses. Never allow children into public rest rooms alone. Send them in groups of threes or fours. Wear comfortable, loose-fitting clothes. Good stout shoes are a must. Have an emergency plan. Inform the children of the time and place for departure. Alert the group to a recall signal. One blast on the whistle means the group has 5 minutes to assemble at the bus; two blasts — 3 minutes; and three blasts — stragglers are on their own, the bus is leaving without them. Have children buddy up. Keep a head count. You should not return with fewer or more students than you started with.

Bring a list of telephone numbers with you. This list should include each student's home phone, the principal's number, and the number of the nearest state police station. When there are problems everyone needs to be informed. If the trip extends beyond the normal day, arrange to have the children dropped off at their home bus stops or arrange a time for parents to pick up their children.

Reasons why some field trips fail

- Adequate preplanning is not carried out — administrative approval, parental approval, site approval, transportation, familiarization with the route, parking, facilities, equipment, prior discussion of the trip.
- The teacher is unenthusiastic. (The teacher's attitude sets the tone for the field trip.)
- The students may have been there before.
- The students may have been forewarned about the kind of field trips you undertake.
- The students may not have had a part in the planning and the selection of the field trip area.
- The objectives of the field trip may be beyond the level of achievement of the students.
- The field trip is taken merely to discharge a duty.
- It is a "Look-see, don't touch, nothing to do" field trip.
- The field trip is overstructured. Everyone is too busy collecting, inspecting, and cataloging. No free time is allotted for individual exploration and investigation.

— The cost of transportation and meals is excessive. Too much pocket money is required.
— An undisciplined student can spoil the trip for most students.
— The trip is too time-consuming. The investment is not worth the return.

Reasons why some field trips succeed

— Careful and meticulous but flexible planning is done by the students and the teacher.
— Enthusiasm is exhibited by the teacher. (Enthusiasm is infectious.)
— A variety of field trips are initiated. (Micro- as well as macro-field trips are possible.)
— Involvement of parents enhances the chances of success. (Field tripping is not necessarily a do-it-yourself endeavor.)
— Pleasurable experiences are programmed into the trip: photos of students involved in various activities; surprise treats (a treasure hunt, hot chocolate at the end of the trail).
— Teachers forget they are teachers and join the ranks.
— The field trip has a purpose.
— A review of the results of the field trip is made to reinforce achievement of the objectives; a review of slides is made; both reviews lay the groundwork for the next field trip.

points to ponder

Design a field trip for your class without leaving the classroom.

Prepare a set of activity cards that will allow each student to investigate an individual problem on a field trip.

Prepare a list of reasons why we shouldn't take field trips. Next, prepare a list of ways to overcome the problems listed.

DISCIPLINE AS A PART OF CLASSROOM MANAGEMENT

Beginning teachers seem to agree that establishing and maintaining order in the classroom is both an awesome responsibility and a challenging task. In our conversations with teachers and prospective teachers, we notice that the word *discipline* means different things to people. It is, perhaps, most frequently used to mean "the degree of control or order that is found in a classroom." To some, it is the set of techniques used to bring about that order. In some contexts, it can mean "self-control" — the extent to which an individual controls personal activities. We also note that the word *discipline* is commonly used to mean "punishment."

Creative-sciencing teachers know that along with the tasks of preparing lessons and devising suitable teaching methods, they must create and maintain a climate for learning. The optimum environment for learning is one in which teachers and students are all actively involved in a learning process relatively free from interruptions.

Many teachers point out that a bewildering variety of events can trigger disruptions that bring learning activities to a halt — anything from an announcement made over the loud speaker to a student's discovery of a fly on the handle of the pencil sharpener. More generally, disorders occur at the beginning and the end of something (and sometimes in between). Interruptions can occur at the beginning of the school day, the class period, the week, the holiday or vacation period, the lunch period, and, of course, during preparations for science time. Invariably, disturbances accompany landmark events such as report card day, awards day, fire drills, substitute teacher day, an accident, and so on. Disturbances such as these are considered normal and children who cause them are also considered normal.

The number and duration of the interruptions that you allow is related to your personality, philosophy, and tolerance. Some teachers permit more interruptions and for longer periods of time than others. Your tolerance for interruptions will vary. What you may view as a discipline problem one day, might cause you to chuckle on another day. Your reactions will depend on many things — your health, how well rested you are, the weather, the season, how well prepared you are, or the announcement by the principal that your pay raise was approved. Normal interruptions should be viewed for what they are. They should be tolerated but never encouraged. Above all, they should be viewed with common sense. It helps if you are consistent in handling these disorders. Children become confused by teachers' inconsistency in reacting to disorderly behavior.

Up to now, we have been talking about the minor disruptions that occur on a day-to-day basis. It is possible that you may face more serious problems such as cheating, lying, stealing, chronic misbehavior, persistent absenteeism, physical aggression, open hostility, destructiveness, and sexual offenses. Such problems are of major concern to teachers and other school officials and should be handled by trained personnel such as the school psychologist, counselor, or principal.

The prevention and defusing of behavior problems

— Operating on the assumption that it is easier to give something up than to recapture it, start the year being firm and, if warranted, loosen up as the year progresses. The reverse procedure is much more difficult.
— Establish the ground rules for behavior you will accept and will not accept. Be sure the ground rules are reasonable and that they can be met. Do not threaten. React expeditiously and confidently. You are in charge. Let the children know it. Let them know that you know they know.
— Avoid overfamiliarity with students. Teachers want all their students to love them. This is not always possible; for many of them to like you should be sufficient. Your assignment is to teach and be respected in the process. Good teachers usually get the love and respect they deserve.
— Don't abuse your authority as the disciplinarian. Be firm but be gentle. Ridicule and embarrassment add fuel to the fire of behavior problems. Fairness, discreet correction, consistency, and impartiality are absolute necessities.
— Check with the principal to make sure that your discipline measures are compatible with school policy. Ask the principal what he or she expects of teachers in maintaining appropriate behavior. Also, determine how far the principal or other administrators will go in supporting your efforts in classroom discipline.
— The most valuable assets you have for maintaining control are your vitality, your enthusiasm, and your voice. Start your science lessons with vigor and enthusiasm. Don't waste time. Convey the idea that what is going on is vital and necessary in the learning process of each individual. Develop a voice that is pleasant to listen to, one that commands attention. Cultivate variations in the way you use your voice that permit you to communicate in many different ways: from whispering a ghostly Halloween story to marshalling the class for a mass exit to avoid an oncoming catastrophe.
— Alternate your approaches. Sometimes direct your remarks to the entire class: "Someone is not paying attention!" This can get all the children thinking, Does the teacher mean me? If this approach

does not work, be direct and identify the individuals. More serious disturbances dictate use of the latter method. Never punish the entire class for the misbehavior of one individual or a small group.

— Always strive to teach as though you were being considered for the school's congeniality award. Then, when you are forced to present a firm posture, it will have more impact.

Your role is to teach. Disciplinary actions must be minimized because they hinder the learning process. Through practice, you will develop your own style of handling disruptions. As you develop your repertoire of disciplinary actions, children should grow in self-discipline. Opportunities need to be provided to nurture self-discipline or peer-group discipline. Often it is easier and safer to hold on to the reins of discipline than risk losing control by transferring them. But the children's mastery of self-discipline is well worth the risk.

Many behavior problems can be avoided if lessons are well-planned and interesting. As often as possible, lessons should exhibit a balance of exposition and student involvement and should be devoid of catch-up and wait time. Nebulous assignments, unclear communication of the objectives of the lesson, repetitive teaching methods, and failure to provide for individual differences are pitfalls that creative teachers avoid.

The very nature of science and its direct relationship to daily life provide many concrete learning experiences. In addition, the numerous challenges of and skills developed through problem-solving tend to minimize classroom disruptions. Teachers should ask themselves the following questions *before* beginning their science instruction:

1. What are the most effective ways of organizing my classroom for laboratory work, discussion, and demonstrations?
2. What are some different ways of developing a science lesson?
3. How can I phrase my science questions more effectively?
4. How much should I guide children in science?
5. What are some effective ways of storing and distributing science materials?

Managing the science classroom

The job of the teacher changes as the classroom organization for a science lesson changes. When working with the entire class, in either a science discussion or a demonstration, the teacher guides the activities by asking questions, giving directions, and making suggestions. Your job is to encourage participation by using the children's ideas. In conducting discussions, make sure that the children hear what you and others are saying. Remember that science equipment and materials unique to a new activity usually distract children and should be kept out of sight until they are needed. Many teachers are successful in designating one part of their room for discussions and another part of the room (preferably with a tile floor and near a sink) as the laboratory. Demonstrations should be conducted with the children sitting or standing in an arrangement that enables all of them to see and hear. If the science demonstration is a starting point for additional activities, make sure that your directions are clear and concise.

When children work individually, they can work at desks or in designated areas of the room. If they work in small groups, a few desks or tables can be pushed together to serve as a working surface. A general rule of thumb: as the individual work increases so should the student-to-student interaction. You need to know your children well enough so that you can have students who can teach each other work together. The teacher must constantly be on the move, showing a genuine interest in the children's activities. Rather than towering above the children, the teacher should sit or kneel down at desk level to get a close view of the work being done, to ask and answer questions, and to give encouragement. The teacher must make a point of talking to each child in the group.

There are many ways of conducting a science activity or discussion. A teacher can work with individuals, pairs of children, or larger groups. A science lesson might be planned so that different activities are going on at the same time. When a child successfully completes one activity, he or she may go on to another. Children also enjoy presenting their own demonstrations.

Taking fifteen minutes or more to distribute materials for a science activity is not efficient. Boredom, horseplay, and more serious discipline problems may arise as the children sit and wait. A smooth and efficient distribution system requires planning. Such planning should give the children a role in handing out materials. It should also involve children in collecting and storing of the materials.

The time of the day that science is taught may have a great influence on how the materials are distributed. Sometimes, it is preferable to schedule science just after recess so that the materials can be distributed while most of the children are out of the classroom. Another successful method is to divide the children into groups at the time of the science lesson, and give each group the responsibility for handing out particular materials. While one group gives out the plastic tumblers, for example, another group distributes the seeds. When the science lesson is over, the groups can collect the materials and store each type in a designated spot. The groups should also clean the equipment, the desks, and the containers.

One person in each group can be assigned to go to the central supply table to obtain liquids or other messy materials for the other members of the group. If many liquids are needed, the liquids can be located in different areas of the room and each group can send one member to get one liquid.

If each child needs to use a set of materials several times, the materials can be kept in a plastic bag with the child's name on it. This reduces collection and distribution problems.

The management and discipline you use during science will vary with different children and changing conditions. None of these approaches, however, should discourage any child from developing his or her creative abilities through science. Creative teachers are flexible enough to prevent this from happening in their classrooms.

points to ponder

Studies show that teachers who treat students as individuals have the best discipline. Why don't more teachers use this technique?

Many teachers involve their children in the development of the rules for science class. Develop a plan to accomplish this in your classroom.

WORKING WITH THE INTELLECTUALLY GIFTED CHILD IN SCIENCE

For years educators believed that one need not worry about gifted children. It was thought that they could and would take care of themselves. The classic statement has always been, "They'll learn in spite of us." Evidence tells us something different, however: gifted children may be the most neglected students in the schools today.

Definitions of *gifted* vary. Earlier in this century, we linked the definition solely to IQ, particularly as measured by the Stanford-Binet test. Those children whose scores on this test placed them in the top 1 or 2 percent were declared to be gifted. Today, we tend to use the definition given by Sidney Marland, the former United States Commissioner of Education.

Gifted and talented children are those . . . who by virtue of outstanding abilities are capable of high performance. These . . . children . . . require differentiated educational programs and/or services beyond those normally provided by the regular school program in order to realize their (potential) contribution to self and society.

Children capable of high performance include those with demonstrated achievement and/or potential ability in any of the following areas:

1. General intellectual ability
2. Specific academic aptitude
3. Creative or productive thinking
4. Leadership ability
5. Visual and performing arts
6. Psychomotor ability[2]

This definition is an attempt to recognize children who possess a variety of talents. The term *talented* generally refers to a specific dimension of skill; for example, in music or art. In most gifted children a strong relationship is found between their giftedness and talented performance. Even though we acknowledge that superior intelligence is only one factor in determining an individual's ultimate success, achievement, or contribution to society, it still remains a basic attribute of the gifted.

It is possible to list some characteristics of the intellectually gifted:

— an outstanding memory
— an early attraction to problems (Gifted children are concerned with the *why* of everything.)
— an early concern with one's own thoughts and feelings, and an ability to analyze situations
— a fascination with reading (An early interest in books often stimulates gifted children to teach themselves to read long before they go to school.)
— an unusually precocious talent in art and music

2. Sidney Marland. *Education of the Gifted and Talented* (Report to the Subcommittee on Education, Committee on Labor and Public Welfare, U. S. Senate. Washington, D.C.: U.S. Government Printing Office, 1972).

— a tendency to seek out and associate with older children and adults rather than children their own age

The direction that a gifted intellect takes depends on many factors, including experience, motivation, interest, emotional stability, mentors, parental urging, and even chance. In the classroom, many gifted children seem to appreciate almost any change from the routine that has come to characterize schooling. Gifted children seem to thrive on acceleration and show no sign of serious maladjustment because of it.

Gifted children, because of their early maturity in reading, are interested in a variety of topics. From this range of interests, three broad categories evolve — math, history, and science. The ability of gifted children to read, coupled with their attraction to problem-solving situations (wherein they can demonstrate their concern for humane resolutions of today's problems) makes the study of history popular with them. Science, unfortunately, is not taught as a basic subject in all elementary classrooms. As a result, gifted children get few opportunities to engage in what could afford them joy and intellectual stimulation.

In elementary science instruction gifted children need special programs specifically designed for them. Being gifted is, in a manner of speaking, a handicap. Gifted children need an environment for learning as special as any we create for other handicapped children. Anxious to address themselves directly to the problem at hand, they express disdain for the slow-paced accumulation of discrete facts, or the "bits and pieces" of average science programs. Gifted children want to be immersed in a problem and engrossed in the pursuit of the larger aspects of the problem. As a teacher, you can emphasize the concepts and basic principles of the subject matter for the gifted children. You can also lead them to explore *how* information is derived instead of concentrating only on *what* the information is. Gifted children enjoy involvement in special projects that allow for independent work. They also need to interact in groups with other gifted children who have similar interests.

points to ponder

Gifted children are "special" children. How will you identify the gifted children in your classroom? What science activities will you provide for them?

WORKING WITH THE EXCEPTIONAL CHILD — MAINSTREAMING

Nothing is so unequal as the equal treatment of all children.
— Anonymous

Most of us ask, Who are the exceptional children? Numerous attempts have been made to define the term *exceptional children*. Most simply, we could say that exceptional children are those who require special educational services if they are to realize their full potential. These children exhibit some characteristics that educators should know about when faced with the task of identifying these children. To understand who these children are, the following groupings may be helpful:

1. those with mental deviations, including children who are (a) intellectually superior and (b) slow in learning — mentally retarded
2. those with sensory handicaps, including children with (a) auditory impairments and (b) visual impairments
3. those with communication disorders, including children with (a) learning disabilities and (b) speech and language impairments
4. those with behavior disorders, including (a) emotional disturbance and (b) social maladjustment
5. those with multiple and severe handicaps, including various combinations of impairments: cerebral palsy and mental retardation, deaf and blind, as well as severe forms of physical and intellectual disabilities.

In the past, educational theory and practice supported placement of children needing special educational services in special classrooms, excluding them from the mainstream. They were taught by special educators, and neither the children nor their teachers were involved with the regular classroom teachers. Such placement was based on the belief that the educational needs of these children required special methods, smaller classes, and different curricula.

Social considerations, such as the desirability of insulating the handicapped child from the potentially negative reactions of other children, added support to the advocates of the separate, special classroom approach. Children with severe handicaps were often excluded from public schools because of the lack of appropriate programs or the presumption that they could not learn there. Changing theory and practice in special education have come as a result of four powerful forces: educators themselves, parents and other child advocates, the courts, and the policy and legislation of state and federal governments.

Over the past decade, evaluative research and model educational programs have demonstrated that many past policies and practices were not appropriate. This awareness dramatically culminated in far-reaching federal legislation. In 1975, Public Law 94–142 was passed and signed into law. This landmark legislation, known as the Education for All Handicapped Children Act, contains the mandatory provision that every school system in the nation must provide a free, appropriate public education for every child between the ages of three and twenty-one. A unique feature of the law is the emphasis on the regular classroom as the preferred instructional base for all children. A new philosophy or approach to educational programming for handicapped children was clearly set forth, and the principle of the least restrictive environment became part of American education.

As states began implementing the least-restrictive-environment principle, the term *mainstreaming* quickly entered our vocabulary. In practice, mainstreaming means the instructional and social integration of exceptional children with other children. Several significant criteria guide mainstreaming. When the educational needs of the handicapped child can best be served through placement with nonhandicapped children, the regular classroom will be the least restrictive environment. Primary instructional responsibility, then, lies with the regular teacher. When the handicaps are severe, however, assignment is made to special classes or separate programs.

Another section of PL 94–142 that has great implications for creative-sciencing requires that an Individualized Educational Program (IEP) be developed and maintained for each handicapped child. An IEP must include:

— a statement of the child's present level of educational performance. (The creative-sciencing pretest and profile can serve as a model; see Section 6.)
— a statement of annual goals, including short-term instructional objectives (like those provided in the module sample, pages 113–120).
— a statement of the specific special education and related services

to be provided to the child, and the extent to which the child will be able to participate in regular educational programs.

— the projected dates for initiation of services and the anticipated duration of the services;

— appropriate objective criteria and evaluation procedures (such as the evaluation models for the three domains of learning) and schedules for determining, on at least an annual basis, whether the short-term instructional objectives are being achieved.

The law is clear in requiring that an IEP be developed for every handicapped child at the beginning of the school year and that it be reviewed annually. Once a child has been identified as handicapped, a conference must be held within thirty days for the purpose of developing the IEP.

The IEP conference must include a number of participants: a representative of the public agency who is qualified to provide or supervise special education, the child's teacher, one or both of the parents, the child when appropriate, and other individuals at the discretion of the parent or agency.

The IEP represents the primary vehicle for assuring quality education for the handicapped. It provides a base for assessing the child's level of performance and for providing an effective education program.

Since creative-sciencing teachers emphasize skill instruction, problem solving, and individualization for all children, they are able to include handicapped children in their science programs. Creative-sciencing teachers use activities that emphasize the senses of hearing and touch for the visually impaired, and those of sight and touch for the deaf. Many of the activities in *Creative Sciencing: Ideas and Activities for Teachers and Children* are excellent for these purposes.

It is unfortunate that, in the past, most handicapped students were barred from pursuing scientific careers because they were not exposed to problem-solving and skill activities in the elementary grades. Creative-sciencing teachers are beginning to rectify this situation because of their belief that *all* children should learn science and that this need is especially acute for the handicapped child.

General references that may be of use to creative-sciencing teachers include the following (from the Council for Exceptional Children, 1920 Association Drive, Reston, Virginia 22091):

SAMPLE INDIVIDUAL EDUCATION PROGRAM

Student _____ Date _____ Teacher(s) _____

School _____ Program _____

1. Present level of performance — academic and nonacademic (List test given and observations made to indicate levels)

Educational strengths: _____ Educational weaknesses: _____

2. Placement decisions

Educational programs and services	Person responsible	Initiation date	Ending date	Percent of time

3. Long-term objectives:

Activities and
materials used

Evaluation

Date begun

Date

Results

Recommendations for
specific procedures/
techniques, materials,
learning styles

Dates of meetings

Committee members pres-
ent; parents must sign and
date

1. *Mainstream Currents*, edited by Grace J. Warfield
2. *Mainstreaming: Educable Mentally Retarded Children in Regular Classes* by Jack W. Birch
3. *Making It Work: Practical Ideas for Integrating Exceptional Children into Regular Classes*, edited by Barbara Aiello
4. *Teacher, Please Don't Close The Door: The Exceptional Child in the Mainstream*, edited by June B. Jordan

Science programs for the handicapped are available from:

1. Biological Sciences Curriculum Study, PO Box 930, Boulder, CO 80302. Titles: *Me Now, Me and My Environment*, and *Me In the Future*.
2. Lawrence Hall of Science, University of California, Berkeley, CA 94720. Title: *Adapting Materials for the Blind*.
3. Chemistry Department, American University, Washington, DC 20009. Title: *Science for the Handicapped*.

The general effect of PL 94–142 will be an increased recognition of the importance of good teaching done in a creative manner. Instruction for all students will become more individualized, and educators will begin to realize that all children are unique and require individualized, specialized programs. Thus, while all students will be educated in the mainstream, that stream will be broader and deeper than in the past, resulting in more educational opportunities for all children.

point to ponder

If the individual education of each child is special, why isn't the individual education of each teacher special?

COMPETENCY TESTING

Most people would agree that after more than twelve years of schooling at a yearly cost exceeding $1800 for each child, a high school graduate should be able to read and write. We are, however, continually amazed by newspaper headlines that tell us about high school graduates suing school systems because they are illiterate. Colleges are being forced to give more remedial reading and writing courses. Publishers are having college textbooks rewritten at the ninth- or tenth-grade reading level.

Minimum competency testing has appeared on the scene. Over three-quarters of the states have passed laws or regulations requiring minimum competency testing. Two-thirds of these states require that tests be given, and nearly one-half require passing such tests as a condition for high school graduation. Furthermore, competency testing is sifting down to lower and lower grade levels. In some states, elementary school children must pass approved tests before they can be promoted. Many states require minimum competency tests before children can be promoted to grade seven.

Most competency testing is being done in reading and mathematics, although some states test all the basic skills. Creative-sciencing teachers recognize that there are certain skills needed for lifelong learning and they assist *each* student in acquiring these skills. While we refer to skills such as observation, inference, measurement, formulating hypotheses, and communicating as sciencing skills, they are, in reality, the minimum competency skills needed for lifelong learning. Teaching these skills will assist students in meeting the minimum competencies required by most states in reading and mathematics.

points to ponder

What competencies should a teacher test for? How do you avoid minimum competencies becoming maximum competencies? Should minimum competency test scores be used to evaluate teachers? Why or why not?

How can pre- and postassessments along with a student profile meet the needs perceived by minimum competency testing?

Cite the pros and cons of a minimum competency testing program in the elementary school, secondary school, and college.

CONTROVERSIAL ISSUES IN SCIENCE

Creative-sciencing teachers do not skirt the ethical issues of science. They realize that education can help children recognize and identify dangerous situations and can help prepare students to deal with them. Rather than simply presenting one side of an issue or just its emotional aspects, creative-sciencing teachers begin with the issue and have the children present as many sides of the issue as possible. Creative-sciencing teachers do not choose sides; they allow children to make decisions so that when they are confronted with real life, they will make wise judgments. Following are some examples of how controversial issues can be dealt with.

Issue 1: smoking

Closeup

1. The average life expectancy of a man of 25 (nonsmoker) is six-and-one-half years greater than that of a man who smokes one or more packs a day.
2. The death rate from lung cancer is ten times greater for cigarette smokers than it is for nonsmokers.
3. Cigarette smoking is a habit among a minority of adults and even a smaller minority of teenagers.

points to ponder

Will you smoke? Why or why not? Does smoking make you more popular? Why or why not? Does smoking make you more sophisticated? How expensive is smoking? Do nonsmokers have the right to stop smokers from smoking in public places?

Additional information for Issue 1 can be obtained from the National Tobacco Institute, Raleigh, NC 27614, and the American Cancer Society, 219 East 42nd Street, New York, NY 10017.

Issue 2: drugs

Closeup

1. Drug use is increasing among younger children (although it varies widely from one school to another).
2. No single cause or set of conditions clearly leads to drug dependency; it occurs in all social and economic classes.
3. Most states have laws requiring instruction in drug education, but most instruction occurs too late in a child's life.
4. Narcotic drugs include some of the most valuable medicines known as well as some of the most abused.
5. A drug is a medicine that helps to make sick people well, but it can also make well people sick.

What They Neglected to Tell Me about . . . / 186

6. Never take drugs that belong to someone else. They may make you sicker.
7. Some things look like drugs, but they are actually poison. They are marked with either the sign of the skull and crossbones — old method — or the snake (SIOP) — new method.
8. Drugs are for sick people.

points to ponder

Cite alternative ways of getting "high" that are not self-destructive. State wise uses of drugs and problems in the use of over-the-counter drugs.

Additional information for Issue 2 is available from: Drug Enforcement Administration, Department of Justice, Washington, DC 20537; The Drug Abuse Council, Inc., 1828 L Street, NW, Washington, DC 20036; and National Institute on Alcohol Abuse and Alcoholism, 5600 Fishers Lane, Rockville, MD 20852.

Issue 3: death

Closeup
1. Death is one part of the cycle of living things.
2. Death is *not* the same as going to sleep.
3. Funerals are mechanisms used by society to cope with death.

points to ponder

What happens when a loved one dies? Have you ever had a pet die? How do burial customs in different countries of the world compare?

Additional information for Issue 3 is available in these articles: W. L. Yarber, "Where's Johnny Today? — Explaining the Death of a Classmate," *Health Education*, vol. 8 (January-February 1977), pp. 25–26; W. L. Yarber, "Death Education — A Living Issue," *Science Teacher*, vol. 43 (October 1976), pp. 21–23.

Issue 4: sexuality

Closeup
1. An understanding of basic reproduction is vital for the continuation of our species.

2. A misunderstanding of sexuality leads to major problems in the intermediate grades including sexually transmitted diseases, pregnancy, birth control problems, and negative feelings about sexuality.

points to ponder

Why are there variations in growth of secondary sex characteristics in children (breast development, underarm hair, menstrual cycle)?

What are some misconceptions about reproduction, nocturnal emissions, and masturbation?

What physical and emotional problems are associated with sexual development in children?

What problems are associated with increased recognition and interest in the opposite sex? Do the changing roles in dating and family life result in confusion for adolescents? If so, what problems are encountered?

Additional information for Issue 4 is available from: W. L. Yarber, "Answering Questions about Sex," *Science Teacher*, vol. 44 (March 1977), pp. 20–22; American Association of Sex Educators, Counselors, and Therapists, 5010 Wisconsin Ave., NW, Suite 304, Washington, DC 20015; John J. Bunt and Linda Brower Meeks, *Education for Sexuality: Concepts and Programs for Teaching* (Philadelphia: W. B. Saunders, 1975).

Issue 5: evolution

Closeup
1. Geologic evidence suggests that organisms change through time.
2. In time, a new species may evolve from an old one.
3. In 1859, Charles Darwin, an English naturalist, published *On the Origin of Species by Means of Natural Selection*, a book which contained a new theory about the evolution of plants and animals.
4. Many fossils have been discovered, but pieces in the puzzle of the history of life are still missing.

points to ponder

Why, more than 100 years after its conception, is the theory of evolution so controversial? What is the difference between a theory and fact? Which animals can adapt to the widest range of environments?

What They Neglected to Tell Me about . . . / 188

Additional information for Issue 5 can be obtained from: *Investigating the Earth* (Boston: Houghton Mifflin Co.); Biological Sciences Curriculum Study, PO Box 930, Boulder, CO 80302.

Clarifying values

All of the issues previously mentioned, and you can think of others, relate to values and their clarification. Children have certain experiences as they grow and learn. From these experiences certain general guides to behavior will evolve. These guides tend to give direction to life and may be called values. Our values show what we tend to do with our time and energy.

Choosing values involves seven requirements in three categories.

Choosing a value
  1. It must be the result of free choice.
  2. It must be chosen from among alternatives.
  3. It must be chosen after thoughtful consideration of the consequences of each alternative.

Prizing a value
  4. We must be happy with the choice.
  5. We must publicly affirm our choice.

Acting with our value
  6. We must do something with the choice throughout life — read about it, form new friendships, or spend money on the choice we value.
  7. We must use the value in several different situations and at several different times.

Through these steps, we define the values that guide us.

Values clarification activities that creative-sciencing teachers use fall into several categories, including:

  1. Rank Ordering: Present a situation and three alternatives. Have each of them ranked from 1 to 3 by each child. Then discuss the rankings in an open and nonjudgmental fashion. Here is an example:

Controversial Issues in Science / 189

If you had fifteen million dollars to spend, how would you rank the following as ways to spend it?
— Clean up polluted rivers and streams.
— Develop a new form of energy for fuel.
— Send an astronaut to Mars.

2. Values Continuum: Copy the illustration below on the blackboard. Note the continuum line drawn between the polarized positions on the smoking issue. Suggest that each student mark on the continuum line where he or she stands on the issue. Then discuss choices. Select another issue that can be polarized and repeat the procedure.

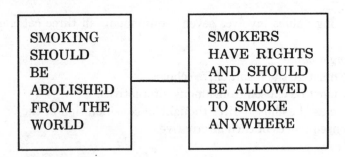

3. I'm Proud: Have each student tell you what he or she is proud of. Suggest a variety of situations such as:

— What did you do in science class today that you are proud of?
— What have you done recently for a friend that you are proud of?
— Which experiment that you have completed are you proud of?

4. I'm Successful: Have each child prepare a Success Chart on which each tells how he or she helped somebody, learned something new, gained like and respect from people, did something his or her family was proud of.

5. My Strengths: Have children tell each other what their strengths are, and then have each child prepare a list of his or her strengths.

6. I Think for Myself: Have children list times when other people decide for them, times when they are allowed to think for themselves, and times when they *must* think for themselves. Then ask them, "When are you allowed to think for yourself in school?"

What They Neglected to Tell Me about ... / 190

Write down some feelings or needs of others to think about when thinking for yourself.

points to ponder

What accomplishments have you made in your teaching career that you are proud of?

What are your science teaching strengths?

Select an issue of concern to you and place it on the value line. Ask your classmates to do likewise.

For further information on values clarification try: Sidney B. Simon, Leland W. Howe, and Howard Kirschenbaum, *Clarifying Values: A Handbook of Practical Strategies for Teachers and Students* (New York: Hart, 1972).

USING MUSEUMS AND RESOURCE PEOPLE

Science centers and museums, in general, provide a whole new field of self-motivating experiences in learning because they offer environmental exhibits that appeal to the senses, emotions, and intellect. They are among the most rapidly developing institutions of learning today. Science museums have more visitors than any other single type of museum. Visitors come to science centers not only to learn facts, but to experience new and interesting phenomena. Most science centers use self-selected experiences that foster the development of logical thinking for various age groups. In addition the museum is one of the few places left where the family can go as a unit.

The science center and museum usually provide experiences that are not available elsewhere. These experiences permit people to keep abreast of new developments in science and technology. Furthermore, these institutions present information on current issues objectively. Their goal is to present enough facts to enable people to make intelligent decisions independently.

Creative-sciencing teachers should visit a museum or science center long before they plan to take their classes. By doing so, they can decide what aspects of the tour they want to emphasize, how they will introduce the tour to children, what arrangements to make for a tour guide (docent), if desired, and what follow-up activities to use after the museum visit. If a class visit is impossible, teachers can prepare a presentation with materials such as slides, tapes, photographs, and brochures to share the visit with their classes.

Most museums respond to the wishes of the classroom teacher. Creative-sciencing teachers request inquiry tours with docents trained in inquiry techniques. Inquiry tours have the following characteristics:

1. They allow the students rather than the docent to draw conclusions.
2. The docent is not afraid to say "I don't know" when presented with questions he or she can not or chooses not to answer.
3. They make students feel that their observations are good and that they should elaborate on them.
4. The docent is not too quick to give the answers. The answers are allowed to evolve from the questions and discussion.
5. They allow for creative discussion by using open-ended and higher-level questions.
6. The docent is not too eager to tell all that he or she knows about an exhibit or object; rather he or she encourages students' thoughts and questions.

Resource people can provide excellent input into the creative-sciencing classroom. Every classroom has children whose parents and acquaintances have interests, hobbies, or occupations which may interest the students. Creative-sciencing teachers capitalize on this by finding out what talents are available through parent-teacher meetings and questionnaires the children fill out. Preparation and planning are important. It is up to the teacher to insure that a resource person is able to communicate with the children at their level.

point to ponder

Why do most museum tours concentrate on giving information rather than on responding to children's interests?

SELECTING SCIENCE BOOKS
FOR CLASSROOM AND LIBRARY

How are books selected for the classroom? The school library? Some states have statewide textbook adoptions, while in others books are purchased at the local level with no state restrictions or limitations. States which have a statewide adoption program usually buy books in cycles — books for each subject are adopted for a five-year period. A science textbook series adopted in 1980 would not be changed until 1985. Other states allow the individual school districts to adopt a science series as needed. With increasing costs, however, more and more schools are reluctant to adopt a new science program frequently.

How should the science textbook be selected? Should the selection be based upon science content, the size of the type, colored pictures, good binding, student guides or supplementary material?

Creative-sciencing teachers first select a textbook or program that emphasizes the thinking abilities of children and is action oriented. Second, they look at the content to find out if it is suitable; third, they examine the layout or format of the book to determine if it is compatible with appropriate pedagogical practices.

Some questions to ask include:

1. Is science presented as inquiry? Is science presented as a structured and directed way of asking and answering questions? Does the material insure that the child will learn that science is not memory or magic, but rather a disciplined way of satisfying human curiosity?
2. Do the materials promote a positive attitude toward science?
3. Do the materials lend themselves to the development of statements of problems for children to solve? Are children given opportunities to observe and compare phenomena, build systems of classification, use the instruments of science, make measurements, conduct experiments, evaluate evidence, draw conclusions, and invent a model or theory?
4. It is difficult to forecast with precision what scientific content the child needs to know; therefore, a science program should emphasize those creative-sciencing skills that will allow the child to pursue knowledge in later life.
5. Do the materials allow for integration with other subjects such as language arts, mathematics, social studies, art, and music?

Whenever possible, the creative-sciencing teacher assists the school librarian or media specialist with the selection of children's books and other learning materials to add to the science section. Monthly reviews of most new children's science books as well as a selection of the best books for each year are published in *Science and Children* (preschool–grade 8) and *The Science Teacher* (grades 7–12). Both are publications of the National Science Teacher Association. Another outstanding review source is *Science Booklist*, published by the American Association for the Advancement of Science (AAAS).

Creative-sciencing teachers recognize that the best evaluators of some aspects of science materials are children. They involve children in the evaluation process leading to the selection of textbooks and other instructional aids. In a creative learning atmosphere, children are free to make recommendations and to critically evaluate the selection and use of materials.

point to ponder

Why do most textbook evaluation forms concentrate on the physical aspects of the book rather than the questions in the list above?

KEEPING CURRENT IN SCIENCE EDUCATION

The creative-sciencing teacher is concerned about keeping current. There are several ways to keep up in science education. You can:

1. subscribe to educational journals such as *Science and Children* or *The Science Teacher* and adapt activities from these journals.
2. receive free educational materials from governmental agencies (such as the Department of Energy), businesses, and industries by writing and asking to be placed on their mailing lists.
3. scan newspaper articles and magazines for topics that will appeal to your students.
4. attend local, state, regional, and national meetings of organizations such as the National Science Teachers Association and the National Association for Gifted Children to obtain new ideas and to look for new materials.
5. volunteer to work with a publisher, college, or university to test new materials in your classroom.
6. enroll in workshops or college courses that apply directly to the science classroom.
7. participate in National Science Foundation (or similar governmental programs) institutes, seminars, and workshops

(usually for credit with your expenses paid by that agency). Contact the appropriate governmental agency for a list of schools you can write to.

8. participate in in-service science programs at your school.
9. ask the children to bring in science related materials and activities of interest to them so that they can share them with you and the class.

Many government sponsored workshops, conferences, and institutes are available for teachers each year. Often these programs allow teachers to obtain graduate credit at a nearby college or university. Tuition, travel, and meal allowances are paid for by the program. For information write to:

— The National Science Foundation, Washington, DC 20550
— The Department of Energy, Washington, DC 20550
— Office of Metric Education, Department of Education, Washington, DC 20550
— Office of Environmental Education, Department of Education, Washington, DC 20550
— The Department of Education, Washington, DC 20550

They will send you a list of colleges and universities that you can write to directly for application information.

points to ponder

Why don't more teachers feel the need to stay current in science? What suggestions can you make to overcome this problem?

Why aren't more college courses offered that help teachers stay current in science? How can this situation be rectified?

PAUSE FOR A SUMMARY

Creative-sciencing teachers frequently mention the following neglected areas in their preparation as teachers: the first day of school; the classroom demonstration; how to order and obtain materials; acquiring, organizing, distributing, and storing science materials; safety in science; field tripping; discipline as a part of classroom management; working with gifted children; mainstreaming; competency testing; controversial issues (smoking, drugs, death education, sex education, values, and evolution); using museums and resource people; selecting science books and other instructional aids for the classroom and library; and keeping current in science education.

— The first day of school can be one of the most important days for both the teacher and the students. Creative-sciencing teachers prepare for the first day by incorporating exciting activities that involve students.
— A classroom demonstration is a controlled performance, usually presented by the classroom instructor, to convey some preselected phenomena. The outcome is generally known in advance.
— Beginning teachers almost unanimously agree that the most difficult task of teaching is that of establishing and maintaining order in the classroom.
— Creative-sciencing activities provide gifted children with joy and intellectual stimulation.
— Creative-sciencing teachers emphasize skill instruction (observing, inferring, classification) with handicapped children in a mainstream environment.
— Creative-sciencing teachers recognize that there are certain skills needed for lifelong learning, and they assist each student in acquiring these skills.
— Creative-sciencing teachers do not skirt the ethical issues of science. They do not choose sides; they allow children to make choices so that when children face issues in real life, they will make wise judgments.
— Creative-sciencing teachers use science centers, museums, and resource people to provide self-motivating experiences in learning.
— Creative-sciencing teachers select science textbook series and/or science programs that emphasize the thinking abilities of children and use action-oriented, hands-on activities.

Creative-sciencing evaluations

Evaluation raises hell with trust!

*The slowest learners in our schools are not the pupils,
they're the educators who run them.*
— Lester Velie

Do all of you remember the story of Peter Rabbit? Here's a new version.

THE TALE OF PETER AND THE RABBIT

"Class, look at this picture and tell me what you see," said the teacher. Hands went up, but the teacher called on Peter, whose hand had not been one of them. "Peter, what is it?"

"It looks like a rat."

The class laughed. Someone said, "Peter is so stupid. He doesn't know a rat from a rabbit."

The teacher said, "Peter, what's the matter with your eyes? Can't you see that it has long ears?"

"Yes," said Peter weakly.

"It is a rabbit, isn't it?"

"Yes," he said.

"Today's story is about a rabbit," said the teacher, pointing to the picture and then the word. "It's a story about a hungry white rabbit. What do you suppose a rabbit eats when he's hungry?"

"Lettuce," said Mary.

"Carrots," said Suzy.

Eugene Grant, "The Tale of Peter and the Rabbit," *Phi Delta Kappan*, vol. 49, no. 3 (November 1967), back cover. © 1967, Phi Delta Kappa, Inc. By permission.

"Meat," said Peter.

The class laughed. Someone said, "Peter is so stupid. He doesn't know what rabbits eat."

"Peter, you know very well that rabbits don't eat meat," said the teacher.

"That depends on how hungry they are," said Peter. "When I'm hungry I'll eat anything my mother gives me, even if I don't like it."

"Don't argue, Peter," said the teacher. "Now class, how does a rabbit's fur feel when you pet him?"

"Soft," said Suzy.

"Silky," said Mary.

"I don't know," said Peter.

"Why?" asked the teacher.

" 'Cause I wouldn't pet one. He might bite me and make me sick like what happened to my little brother the time one got on his bed when he was sleeping."

The class laughed. Someone said, "Peter is fibbing. He knows his mother doesn't allow rabbits in bed."

After the class had read the story and had recess, the teacher said to the supervisor, "I hate to sound prejudiced, but I'm not so sure that this busing from one neighborhood to the other is good for the children."

The supervisor shook his head sadly and said to the teacher, "Your lesson lacked one very important ingredient."

"What was that?" asked the teacher.

"A rabbit," said the supervisor.

SCIENCE TEACHING AND THE THREE DOMAINS OF LEARNING

If we expect to teach the whole child, then we should expect to teach science in all three learning domains (cognitive — content; affective — attitudes; and psychomotor — physical motion). Unfortunately, most instruction in schools deals with only the lowest levels of cognitive instruction (the Terrible Three) at the expense of the higher cognitive levels and with complete neglect of the affective and psychomotor domains.

The role of the teacher is to teach in all three domains of learning.

The cognitive domain (content)

Cognitive learning is accumulating content knowledge and may be classified in six levels: knowledge, comprehension, application, analysis, synthesis, and evaluation.[1] Whenever possible, cognitive learning should focus upon the five higher levels of learning rather than just knowledge or comprehension. To assist you in recognizing the different cognitive levels try classifying these questions using the levels shown on the Cognitive Chart of Action Verbs.

Level   *Question*

___ 1. Name each of the weather map symbols below.

___ 2. Which of the characteristics below are important in describing a rock's texture?

___ 3. Obtain a blank map of Australia from your teacher. Study both the average precipitation map and elevation map shown below. On the basis of these two maps, where do you think that the river systems originate in Australia?

___ 4. Select the profile of the stream with the greatest potential energy.

___ 5. The features on the map are sand dunes. What is the direction of the prevailing winds in this area?

___ 6. Energy exists in many different forms. State three of them.

___ 7. The scale diagram below represents the orbits of Mars and Mercury. The minimum distance between Mars and Mercury is 106 million miles. What is the radius of the Mars orbit?

___ 8. Leap year occurs every fourth year, and then February has 29 days instead of 28 days. What is the purpose of the leap year?

___ 9. Time zones have been established for the earth. What are the advantages and disadvantages of having time zones?

___ 10. Why are geologists interested in the origin of the moon's surface?

After you have classified each question according to cognitive level, check your answers with those on page 202. If your answers are different from ours, try to infer why.

Cognitive Chart of Action Verbs

I. Knowledge — recalling the terminology, specific facts, principles, generalizations, and theories unique to a subject area.

1. Benjamin S. Bloom, ed., *Taxonomy of Educational Objectives, Handbook I: Cognitive Domain* (New York: McKay, 1956).

1. Identifies	4. Lists
2. Names	5. Selects
3. Chooses	6. Distinguishes

II. Comprehension — translating knowledge into one's own thoughts and words thus enabling one to understand relationships.

1. Computes	4. Demonstrates
2. Measures	5. Selects
3. Matches	6. Balances

III. Application — using previously acquired knowledge and understanding to solve problems in new or unique situations.

1. Compares	4. Calibrates
2. Groups	5. Dissects
3. Arranges	6. Operates

IV. Analysis — to reduce ideas to their component parts.

1. Selects hypotheses	4. Limits
2. Estimates	5. Infers
3. Interrelates	6. Reflects

V. Synthesis — to put parts together to make new patterns and to encourage divergent thinking.

1. Proves	4. Predicts
2. Extrapolates	5. Infers
3. Interpolates	6. Deduces

VI. Evaluation — the most complex of all cognitive learning, it requires the combining of the previous five levels.

1. Controls variables	4. Questions
2. Rejects	5. Interprets
3. Verifies	6. Doubts

From Robert B. Sund and Anthony J. Picard, *Behavioral Objectives and Evaluational Measures: Science and Mathematics*, pp. 41–42. © Charles E. Merrill Publishing Company, 1972. Reprinted by permission.

1. Level I — *name* indicates knowledge
2. Level IV (Analysis) — making inferences as to which characteristics are important
3. Level V (Synthesis) — deducing or inferring
4. Level II (Comprehension) — understanding relationships
5. Level III (Application) — solving problems in a different situation
6. Level I (Knowledge) — state or name from memory
7. Level V (Synthesis) — put parts together to make new patterns
8. Level III (Application) — using previous knowledge in new or unique situations
9. Level VI (Evaluation) — making interpretations based upon the five lower levels of learning
10. Level VI (Evaluation) — making interpretations

How well did you do? Try writing your own list of questions that illustrate your ability to recognize all six levels of the cognitive domain. Now let's move on to the affective domain.

The affective domain (attitudes)

The *most important* of the three domains of learning is the affective domain (attitudes). As we have said before, the affective outcomes of school learning should be of primary importance to the teacher. Once a child's attitude is formed, it is very difficult to change. A child's self-concept (the way a child feels about himself) is paramount to good learning experiences. A child spends from ten to sixteen years in school, a total of over 20,000 hours devoted to school and related activities. The average student enjoys few hours in which he or she is not judged by teachers, peers, family, and others. At no other time in his or her life will the student be judged by so many so often. Unfortunately, rather than considering the affective aspects when designing instruction, most textbooks, teachers, and schools focus on the lowest level of how to learn for later life — factual (cognitive) material as opposed to affective learning.

Many studies have shown a relationship between cognitive success and positive or negative attitudes toward school. If a child has a series of positive school learning experiences, then he or she is likely to develop a generally positive attitude about school and learning. If, however, the learning experiences are generally negative and his or her achievement is regarded as inadequate by the student, the teachers, and the parents, the child is likely to develop a negative attitude toward school and learning.

Unfortunately, once a negative attitude has been instilled in a student in the elementary school, it is almost impossible to change in later life. Academic self-concept is clearly defined by the end of the primary grades, and the teacher must bear a major responsibility for assisting the student in acquiring a positive or negative opinion about school and learning.

The four major attitudes in science that are a part of scientific literacy and that can be evaluated are (a) curiosity, (b) inventiveness, (c) critical thinking, and (d) persistence. Here are definitions and examples of various kinds of behaviors associated with the four areas:

(a) *curiosity*

Children who pay particular attention to an object or event and spontaneously wish to learn more about it are being curious. They may give evidence of curiosity by
— using several senses to explore organisms and materials;
— asking questions about objects and events;
— actively investigating energy transfer in various systems;
— showing interest in the outcomes of experiments.

(b) *inventiveness*

Children who generate new ideas are being inventive. These children exhibit original thinking in their interpretations. They may give evidence of inventiveness through verbal statements or actions by
— using equipment in unusual and constructive ways;
— suggesting new experiments;
— describing novel conclusions from their observations.

(c) *critical thinking*

Children who can provide sound reasons for their suggestions, conclusions, and predictions are thinking critically. They may exhibit critical thinking, largely through verbal statements, by
— using evidence to justify their conclusions;
— predicting the outcome of untried experiments;
— justifying their predictions in terms of past experience;
— changing their ideas in response to evidence or logical reasons;
— pointing out contradictions in reports by their classmates;
— investigating the effects of selected variables;
— interpreting observations in terms of the amount of energy transferred.

(d) *persistence*

Children who maintain an active interest in a problem or event for a longer

period than their classmates are being persistent. They are not easily distracted from their activity. They may give evidence of persistence by
— continuing to investigate materials after their novelty has worn off;
— repeating an experiment in spite of apparent failure;
— completing an activity even though their classmates have finished earlier;
— initiating and completing a science project.[2]

Can you identify an affective behavior and classify it? Try classifying each of the ten statements below. Check your answers with the ones provided, and then write ten examples of your own.
Classify each statement:

    a. Curiosity
    b. Inventiveness
    c. Critical Thinking
    d. Persistence
    e. Not an Affective Behavior

— 1. The student asks why mass is recorded as weight.
— 2. The student ignores past experience when making a conclusion concerning whether or not life exists on Mars.
— 3. The student repeats a density experiment after apparent failure.
— 4. The student correctly states Newton's third law.
— 5. The student suggests using a pendulum to make a sand picture.
— 6. A student checks out a thermometer to conduct a change-of-state experiment with ice at home.
— 7. A student suggests an experiment to record the respiration rate of a goldfish.
— 8. The student tests the reaction rate of two different kinds of metals with acid in addition to the required metal.
— 9. After receiving a grade of C on his or her experiment, the student repeats the experiment and submits additional results.
— 10. After listening to an answer from a classmate during a discussion about evolution, a student presents evidence in opposition to the first answer.

2. Reprinted with permission from the *Energy Sources Evaluation Supplement*, written and published by the Science Curriculum Improvement Study. Copyright © 1973 by the Regents of the University of California.

Here are the answers. How well did you do?

1. c	4. e	7. b	10. c
2. c	5. b	8. a	
3. d	6. a	9. d	

The psychomotor domain (physical motion)

The psychomotor domain is concerned with the development of muscular skill and coordination. This domain includes such skills as using tools (hammer, screw driver, wrench), reading rapidly, playing a musical instrument, using a microscope, playing sports. While intellectual skills enter into each of these psychomotor skills, the primary focus is on the development of the psychomotor skill involved. Most successful teachers divide their emphasis among the three domains as follows:

K–3 Teachers
20% Cognitive Instruction
40% Affective Instruction
40% Psychomotor Instruction

4–6 Teachers
40% Cognitive Instruction
30% Affective Instruction
30% Psychomotor Instruction

7–12 Teachers
50% Cognitive Instruction
25% Affective Instruction
25% Psychomotor Instruction

College Teachers
60% Cognitive Instruction
20% Affective Instruction
20% Psychomotor Instruction

point to ponder

How do you plan to divide your instructional time among the three domains? Justify your division.

Science Teaching and the Three Domains of Learning / 205

EVALUATION AND COMPETENCY TESTING
IN THE COGNITIVE DOMAIN

As we turn toward the evaluation of children, we find that creative-sciencing teachers evaluate children using all three domains: cognitive (content), affective (attitude), and psychomotor (physical motion). Creative-sciencing teachers provide an environment that encourages all children to succeed. They realize that evaluation is a continual process and of an individualized nature. They understand that the so-called unsuccessful students (as viewed in schools) *can* be successful in science because these students have many desirable attributes:

— They are very inventive in the use of nonstandard oral language.
— They are able to progress and to be successful when dealing with things that they are interested in — frogs, boats, airplanes, engines, and electricity.
— They are able to concentrate for long periods of time when dealing with things of interest to them.
— They show considerable insight into human behavior.
— They are not as easily frustrated as many straight A students.
— They are apt to frustrate the school system with some highly imaginative approaches.
— They approach and deal with various situations in unusual or unconventional, but appropriate, ways.

As illustrated in our three module examples in Section 3, creative-sciencing teachers preassess children according to a set of performance objectives. This procedure is also called criterion-referenced testing. Then activities are designed, based on the results of the preassessment, to help the child succeed. Success can be measured by postassessment. Many teachers provide children with profiles of the science-process skills so that everyone — the child, the teacher, the parent, and the administrator — can be informed of a child's progress (see the sample profile, page 207). More and more schools are adopting the profile and conference report as a positive alternative to grading. If you, however, must administer grades, a skills profile can provide the necessary information. If the children and you agree that they must achieve a postassessment score of ninety percent of the possible points for an A, your grading scale has been established.

We should point out, however, that since success is our goal, children

SAMPLE PROFILE OF A STUDENT'S SCIENCE-PROCESS SKILLS

Name _____

Section _____

Preassessment date _____

Postassessment date _____

Module number	Science process	Number of points possible	Preassessment score	Postassessment score	Change plus or minus
1	Observing	6			
2	Inferring	8			
3	Measuring	14			
4	Using numbers	10			
5	Classifying	10			
6	Space/time	8			
7	Communicating	12			
8	Interpreting data	8			
9	Controlling variables	10			
10	Experimenting	10			
11	Defining operationally	6			
12	Instructional objectives	8			
	Totals	110			

enrolled in classes taught by creative-sciencing teachers will probably earn more As than ever before.

Test items used by creative-sciencing teachers are not low-level cognitive-knowledge questions such as:

— Draw the Beaufort wind symbol for the wind coming from the northwest at 37 miles per hour.
— Change 35°F to its corresponding Celsius temperature.

Instead their questions, written at a higher level and involving thinking, are like the following:

— Using the materials provided, construct a four-stage classification scheme.
— Given the following data, construct a graph and make two predictions from it, one of which is an extrapolation and one of which is an interpolation.

Write several assessment items of your own. Remember they should involve thinking and allow for a variety of answers. In addition, they should be skill questions related to observing, measuring, classification. A sample science process measure assessment is presented in Appendix A.

EVALUATING YOURSELF

Creative-sciencing teachers evaluate themselves and their instruction *first*, to determine if they are meeting the needs of their students. Questions and statements that creative-sciencing teachers concern themselves with at the end of each day are:

— Write down something everybody learned.
— Write down something nobody learned.
— Write down something different each child learned.
— Name the child who knew everything you taught beforehand.
— Name the child who learned nothing.
— Did you do anything other than mass instruction?
— Did you challenge pupils' intellects?
— Did anyone work on an independent learning problem?
— Did you reject any child?
— How many children failed?

— Did anyone work at the board? Did you?
— Did pupils do anything other than listen to you, write, read, answer your questions?
— Did any child ask a question?
— How often did you say, "O.K." or "All right"?
— Did you repeat every answer?
— How often did each child get a chance to talk, ask, tell, answer?
— Did any child help another?
— Did the class laugh; did you — at someone, something, with someone?
— Were you angry — why?
— Was any child angry — why?
—How many pupils did you praise?
— Did you teach reading — (not just hear children read)?
— Did you teach new words — before or after reading?
— Did you read to the class?
— How do you feel about the day's work? [3]

The reactions to these questions and statements are used as aids by the creative-sciencing teachers in preparing for the next teaching day.

Creative-sciencing teachers encourage evaluation of themselves and their classrooms by their peers, the administration, and by their children. A sample administrator evaluation is shown on the next four pages. The emphasis is upon a child-centered classroom, with many of the traditional evaluation items eliminated.

3. Gertrude Noar, *Individualized Instruction — Every Child a Winner*, pp. 111–112. Copyright © 1972 by John Wiley and Sons, Inc. Reprinted by permission of John Wiley and Sons, Inc.

TEACHER EVALUATION FORM

	OK	Let's talk about it

1. PERSONAL

a. Appearance. _____ _____

b. Personality. _____ _____

c. Acceptance by staff. _____ _____

d. Acceptance by parents. _____ _____

e. Ability to adjust to the positive approach to education. _____ _____

f. Enthusiasm and motivation for the job. _____ _____

g. Warm and sincere with children. _____ _____

h. Can accept the opposite sex and age as a capable teacher. _____ _____

i. Believes in school philosophy of a "school without failure." _____ _____

2. EFFECTIVENESS AS A TEACHER

a. Teacher is friendly and approachable. _____ _____

b. Teacher works at making it pleasant to be a child in his or her room. _____ _____

"Teacher Evaluation Form," Van Buren Elementary School, Plainfield, Indiana. By permission of Eugene DeBaun, Principal.

	OK	Let's talk about it
c. Teacher accepts children's delays and shortcomings without putting them down.	_____	_____
d. Teacher has ability to make every child feel important and wanted.	_____	_____
e. Teacher has children using tape recorders, record players, 8-mm. movie projectors, overhead, and other equipment.	_____	_____
f. Teacher has found more positive ways to teach self-discipline to a child than keeping him in at recess and isolation.	_____	_____
g. Teacher can work equally well with every child.	_____	_____
h. Teacher has ability and enthusiasm to make every subject exciting.	_____	_____
i. Teacher uses areas for evaluation other than grades, achievement, and I.Q. scores.	_____	_____
j. Teacher has children friendly and supportive of teacher, principal, and school.	_____	_____

	OK	Let's talk about it
k. Teacher's classroom shows teacher and children working in cooperative teamwork.	_____	_____
l. Evidence that children are active in decisions affecting them.	_____	_____
m. Communication between children and teacher is accepted and can be openly questioned.	_____	_____
n. Teacher has nonjudgmental class meetings so children can talk.	_____	_____
o. Teacher allows children freedom to fail.	_____	_____
p. Teacher keeps the importance of discipline in its proper place.	_____	_____
q. Teacher allows children to use their own judgment in solving problems.	_____	_____
r. Teacher does not feel he or she is great dispenser of knowledge.	_____	_____
s. Atmosphere in room.	_____	_____

	OK	Let's talk about it
t. **Room decorations and bulletin boards are child centered and child prepared.**	_____	_____

Comments from teachers in conference after evaluation:

EVALUATION IN THE AFFECTIVE DOMAIN

Evaluation in the affective domain by creative-sciencing teachers is on the rise. We have found that a child's attitude toward science and school, in general, becomes more favorable when the creative-sciencing method is used.

The four major attitudes in science that can be evaluated are (a) curiosity, (b) inventiveness, (c) critical thinking, and (d) persistence. The behaviors associated with these four main attitudes have already been stated and will allow for the selection of specific evaluation schemes. In addition, an evaluation of a child's perception of the classroom environment can be accomplished through the use of pictures in the primary grades and a science report card in the intermediate grades (see the samples, pages 214–216).

A child's self-concept and attitude can be positively reinforced through the use of Happy Note forms (page 217). These are designed to be given to children as rewards for positive behaviors and attitudes and to help improve self-concept. These notes have been found to be very successful in promoting positive attitudes.

1• 2: 3⋮

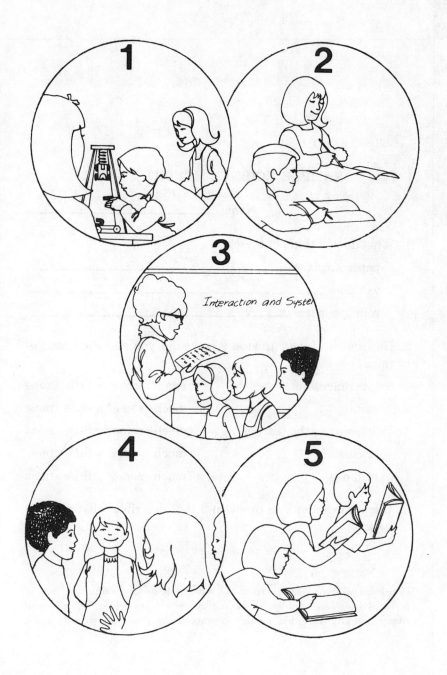

By permission from *Interaction and Systems Evaluation Supplement*, written and published by the Science Curriculum Improvement Study. Copyright © 1972 by the Regents of the University of California.

Evaluation in the Affective Domain / 215

SAMPLE SCIENCE REPORT CARD

Name _____

1. What do you think of these?

   paper airplanes _____

   variables _____

   rolling spheres _____

2. How much of your time in science did you spend on each of these?

experimenting	much	some	a little	none
writing	much	some	a little	none
listening to the teacher	much	some	a little	none
discussing	much	some	a little	none
reading	much	some	a little	none

3. Circle the activity in question 2 that describes what you like best.

A HAPPY NOTE

Good news from school !

to _____

from _____

_____

_____

_____

_____

_____

_____

A good affective evaluation includes unfinished thinking problems such as:

I took a trip with my family to the zoo. It was big and noisy and crowded. It was fun too, until I got separated from my parents. I looked around and they were gone.
What would you do?
Write an ending for the story.
How would you know if your decision was right or wrong?

The value continuum or line is also excellent for affective evaluation. On the blackboard, draw a line with a large circle on each end:

Then choose a two-sided issue; the greater the polarity the better. Present the issue to the students and explain the two opposing views. Write a one-word summary of one view in each circle. Next, present two other views that lie in between the extremes. Mark these on the continuum line. Express your own view on the issue and mark it on the line too. Have each child decide his or her position on the issue and mark it on the line. Finally, have the children discuss their views on the issue.

As an alternative to the blackboard sketch, you can chalk the circles on the floor and tape a line between them. Follow the procedure above, but have the children move to the place on the tape continuum that represents their views on the issue. Try these opposing points of view for starters:

— Conserve energy *vs.* Use energy
— Save our environment *vs.* Forget our environment
— Science is moral *vs.* Science is immoral
— Evolution *vs.* Nonevolution
— It's OK to use drugs *vs.* It is not OK to use drugs

Reaction situations like the next one also work well for evaluative purposes. Remember, there can be more than one correct response.

The teacher assigns a science project to each class member. Your father does most of the project, but you turn it in and say you did the whole project yourself. The teacher asks you for an explanation because you couldn't have done it yourself.

What is your reaction?
Is it positive or negative?
What other response could you have made?
Is this reaction positive or negative?

Now write some of your own reaction situations.

EVALUATION IN THE PSYCHOMOTOR DOMAIN

Children involved in creative-sciencing activities are doing things — hammering, sawing, drilling, and soldering. These are a few of the psychomotor skills necessary for a successful sciencing experience. Activities should be designed to give children experiences in soldering two wires together before they make a circuit board; drilling a hole before they build a boat; and using tin snips before building animal homes. *Creative Sciencing: Ideas and Activities for Teachers and Children* contains several psychomotor activities that could be used for teaching and evaluative purposes, among them soldering and cutting wood with a jigsaw.

One psychomotor evaluation utilizes the skill of measuring and deals with how students change through the school year and into the future. Have each child record the following information in the fall, winter, and spring: height, weight, hair color, eye color, sex, and one measurement of your choosing. Which information changed? Why? Which information stayed the same? Why? Next, have the students recall this information about themselves as they were three years ago and then predict how this information will be different three years from now and six years from now.

In conclusion, evaluation is a serious segment of the educational process. We can use evaluation as a learning experience or as a bludgeon. Much evaluation presupposes that what is being taught is desired by the learner. When, as educators, we commit ourselves to teaching what is relevant, and when we present this learning as deeply rooted in experience, and present it in an open dialogue, based on mutual trust and respect between student and teacher, then good education is taking place.

BECOMING A CREATIVE SCIENCER

As you begin involving children in creative-sciencing experiences, you will find that you can spend some time observing your students in action. Here are some questions to consider in finding out how successful your creative-sciencing program is becoming:

— Do the students talk with each other about their work?
— Do they initiate activities which are new to the classroom?
— Do they persist over a period of days, weeks, or months on things which capture their interest?
— Do they have real interests of their own?
— Are they able to say, "I don't know," with the expectations that they are going to do something about finding out?
— Do they exhibit any initiative, have they developed any skill, in finding out what they want to know?
— Do they continue to wonder?
— Can they deal with differences of opinion or differences in results on a reasonably objective basis, without being completely swayed by considerations of social status?
— Are they capable of intense involvement? Have they ever had a passionate commitment to anything?
— Do they have a sense of humor which can find expression in relation to things which are important to them?
— Do they continue to explore things which are not assigned — outside of school as well as within?
— Can they afford to make mistakes freely and profit from them?
— Do they reflect upon their errors and learn from them?
— Do they challenge ideas and interpretations with the purpose of reaching deeper understandings?
— Are they charitable and open in dealing with ideas with which they do not agree?
— Can they listen to each other?
— Are they willing to attempt to express ideas about which they have only a vague and intuitive awareness?
— Are they able to make connections between things which seem superficially unrelated?
— Are they flexible in problem solving?
— Are they willing to argue with others?
— Can they suspend judgment?
— Are they capable of experiencing freshly and vividly?
— Do they know how to get help when they need it and to refuse help when appropriate?

— Are they self-propelling?
— Can they accept guidance without having to have things prescribed?
— Are they stubborn about holding on to views which are not popular?
— Are they inner-directed?
— Can they deal with distractions, avoid being at the mercy of the environment?
— Are they intellectually responsible?
— Do they recognize conflicting evidence or conflicting points of view?
— Do they recognize their own potential in growing towards competence? [4]

PAUSE FOR A SUMMARY

— If we expect to teach the whole child, then we should expect to teach and evaluate science in all three learning domains — cognitive (content), affective (attitudes), and psychomotor (physical motion).
— Cognitive learning is accumulating content knowledge and may be classified in six levels.
— The most important of the three domains of learning is the affective domain (attitudes).
— The psychomotor domain is concerned with the development of muscular skill and coordination.
— Creative-sciencing teachers preassess children according to a set of performance objectives (criterion-referenced tests). They also provide each child with a profile of his or her science-process skills so that everyone can be informed of the child's progress.
— Creative-sciencing teachers evaluate themselves and their instruction first, to determine if they are meeting the needs of their students.
— Creative-sciencing teachers encourage evaluation of themselves and their classrooms by their peers, by the administration, and by the children.

4. From William P. Hull, "Things to Think About While Observing," *The ESS Reader* (Newton, Mass.: Educational Development Center, 1970), pp. 153–154. By permission.

Appendix

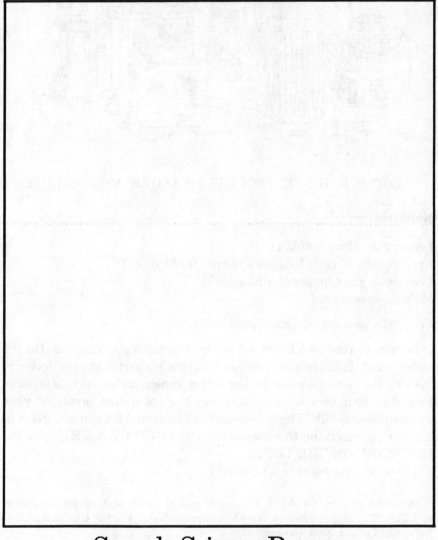

Sample Science Process
Measure Assessment

SAMPLE SCIENCE PROCESS MEASURE ASSESSMENT

Materials needed

1 sugar cube (for Module 1)
1 yellow cube (2 cm³) for each student (for Module 3)
Centimeter graph paper (for Module 7)
Multiple answer card

For profile used see Section 6, page 207.

Be sure to read and follow all of the directions given to you. On the answer card, indicate your answer selection by darkening the letter in the bracket corresponding to the letter choice under each test item. Please do your own work so that you can get a true profile of your science-process skills. The assessment will be scored for you and you will not receive a grade on this evaluation. DO YOUR OWN THING!!! DO NOT MARK ON THE TEST!!

(Note: correct answers are circled.)

Sample developed by Gerald H. Krockover and Marshall D. Malcolm for Education 323, "Teaching Science in the Elementary School," Purdue University, West Lafayette, Indiana, 1969. Revised 1974.

MODULE 1
OBSERVING

Use the object given to you and do the following three items.

1. Select the *best* description of the object.

    A. Cuboidal, candy smell, rough, white, loud sound when dropped

    B. Cuboidal, hard, rough, sweet, white

    C. White, loud sound when dropped, candy smell, rough, sweet

    D. White, rough, candy smell, sweet, hard

2. Select the *best* description of the object.

    A. Cuboidal, 13 mm on a side

    B. Cuboidal small mass

    C. White cuboidal mass

    D. White small mass

3. Select the *best* description of the object.

    A. Lightweight, hard, white

    B. Rough, sweet, white

    C. Sweet, hard, white

    D. Sweet, white, breaks when dropped

MODULE 2
INFERRING

Read the following paragraph and answer the next three test items:

Assume that the room in which you are sitting has no windows. Assume that you had previously been outdoors (5 or 10 minutes ago). You noticed that the sky was extremely black and thunder and lightning were evident. However, it was not raining. A few minutes after you are seated, a late student wanders into your classroom. She is carrying a plastic raincoat on her arm and is wearing a rain hat. Her dress is wet all around the bottom, up 6 or 8 inches from the bottom.

4. Which one of the following statements is *not* an inference?

   (A.) Her dress is wet.
   B.   It is raining.
   C.   She has been wearing a raincoat.
   D.   She walked through a puddle of water.

5. If it is raining outside, which one of the following statements would *best* support this fact?

   A.   It is thundering.
   B.   Linda was late to class.
   (C.) The girl's rain hat is wet.
   D.   You see lightning.

Look at your instructor.

6. Which of the following is *not* an observation of your instructor?

   A.   Appears to be a male or female
   B.   Has hair
   (C.) Over forty years old
   D.   Wears clothes

You are given a circuit board and a test-light. A *red* wire leads from one terminal of a battery to one or more of the upper contacts (A, B, C); and a *green* wire leads from the other terminal to one or more of the lower contacts (D, E, F). Using your test-light, you collect the following data:

Contacts	Light glowed?	Contacts	Light glowed?	Contacts	Light glowed?
A-D	No	B-D	No	C-D	No
A-E	Yes	B-E	No	C-E	Yes
A-F	Yes	B-F	No	C-F	Yes

7. Which of the following diagrams would *correspond* to the actual wiring diagram?

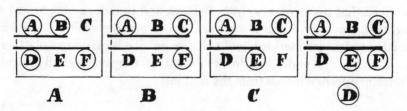

MODULE 3
MEASURING

Go to the measuring station. Using the block provided, determine the following measurements:

8. Length of a side

    A. 1 cm
    (B.) 2 cm
    C. 3 cm
    D. 4 cm

9. Area of any one face

    A. 1 cm²
    B. 2 cm²
    (C.) 4 cm²
    D. 6 cm²

10. Total area of all faces

    A.   6 cm²
    B. 12 cm²
    (C.) 24 cm²
    D. 36 cm²

11. Volume of the block

    A. 1 cm³
    B. 4 cm³
    C. 6 cm³
    (D.) 8 cm³

12. Mass of the block

    A. 4.0 g
    B. 4.5 g
    (C.) 5.0 g
    D. 5.5 g

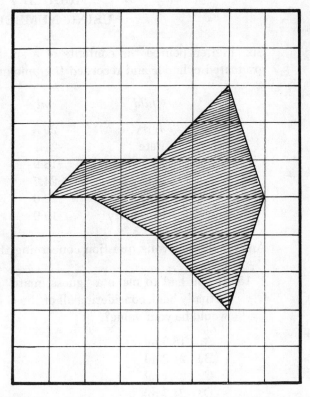

13. Density of the block

    A.   .625
    (B.) .625 g/cm³
    C.   1.6
    D.   1.6 g/cm³

14. What is the area of
    the pattern?

    A.   13.0 cm²
    B.   13.5 cm²
    (C.) 14.5 cm²
    D.   15.4 cm²

MODULE 4
USING NUMBERS

Six children poured the contents of a small container of water into a graduated cylinder and recorded the amount of water in ml as follows:

Child	ml
Jerry	20.3
Pete	21.1
Al	18.9
Hal	24.7
Phil	22.0
Rosemary	19.9

Answer the following question concerning the activity described above:

15. If you had to make a "guesstimate" of how much the container actually held, considering all of the children's measurements, what would be your value?

    A. 18.9 ml
    (B.) 21.2 ml
    C. 22.1 ml
    D. 24.7 ml

Read this problem and answer the following question about the results:

Three Olympic champions of the 800 meter run made these times:
(a) 109.2 seconds, (b) 107.7 seconds, (c) 106.3 seconds.

16. What was the mean rate per 100 meters for the three runners?

    A. 13.29 seconds
    B. 13.46 seconds
    C. 13.51 seconds
    D. 13.65 seconds

17. How much is the volume in the tube?

    A. 12 ml
    B. 14 ml
    C. 16 ml
    D. 18 ml

Answer the following problems:

18. On a recent flight, a jet liner was traveling the 4800 km from Los
Angeles to New York at a ground speed of 1200 km/hr. How long
did it take the plane to make the trip?

    A. 4.00 hours
    B. 4.36 hours
    C. 4.48 hours
    D. 4.80 hours

19. At the end of five days, Nancy, Jenetta, Linda, Tommy, and
Bobby combined the growth of their bean plants. Their total
growth was thirty centimeters. What was the average daily growth
rate for each of the five plants?

    A. 1.0 cm/day
    B. 1.2 cm/day
    C. 5.0 cm/day
    D. 6.0 cm/day

MODULE 5
CLASSIFYING

Observe the objects illustrated on page 231 to answer questions 20
through 24. Use a dichotomous classification system to determine stages.

20. Which of the following is *not* an observable property of the objects illustrated that might be used in classifying the objects into subsets?

    (A.) Color
    B. Margin
    C. Shape
    D. Size

21. How many stages would it take to classify this set of objects into eight separate categories, one category for each object?

    A. 1
    B. 2
    C. 3
    (D.) More than 3

22. What observable property *cannot* be used to divide these objects into various subgroups?

    A. Edges
    (B.) Material
    C. Shape
    D. Size

23. Observe only objects 1, 2, 6, and 7. What is the *least* number of stages needed to separate all of the items into individual categories?

    A. 1
    (B.) 2
    C. 3
    D. More than 3

24. Observe objects 1, 2, 3, and 4. What is the *least* number of stages that would be needed to divide this group of objects into individual categories?

    A. 1
    B. 2
    (C.) 3
    D. More than 3

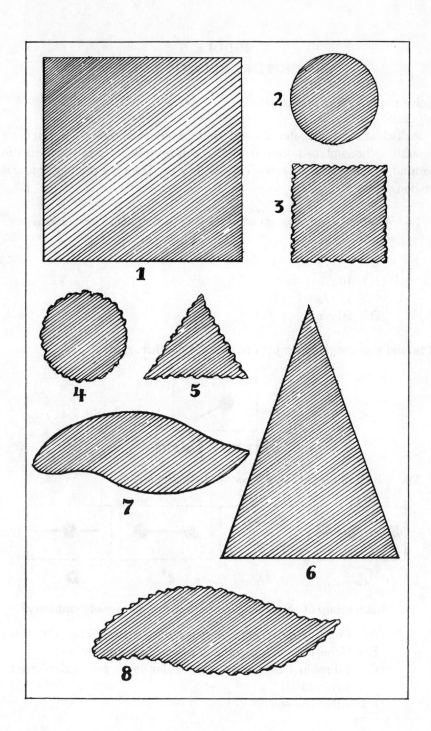

MODULE 6
SPACE/TIME RELATIONSHIPS

Solve the following problem:

A kickball is 0.5 meter around (circumference). From the pitcher's mound to homeplate is ten meters. When a child rolls the ball on the ground, it takes two seconds for the ball to roll from the pitcher's mound to homeplate.

25. What is the *average* speed of the ball in number of revolutions per second (r/s)?

    A.   5 r/s
    B.   10 r/s
    C.   15 r/s
    D.   20 r/s

Pretend you are given a mirror to use on this Mirror Card:

Mirror Card

26. Which of these patterns *cannot* be matched?

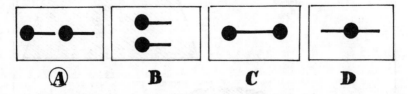

    Ⓐ   B   C   D

27. Which group of objects has *more* than two planes of symmetry?

    A.   Cone, cylinder, sphere, square-based pyramid
    B.   Cube, ellipsoid, rectangular-based pyramid, sphere
    C.   Ellipsoid, hemisphere, rectangular block, triangular-based pyramid
    D.   All of the above

Look at this assortment of shapes:

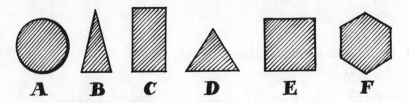

A   B   C   D   E   F

28. Which group of objects has *more* than two axes of symmetry?

    A. A, B, C, F
    B. A, C, D, E
    C. A, D, E, F
    D. B, C, D, E

MODULE 7
COMMUNICATING/PREDICTING

Here is a table of data collected during an experiment to see how long it took a white rat to travel a maze on successive days.

Day	Time (in seconds)
1	120.2
2	89.8
3	60.7
4	41.3
5	34.2
6	25.4

Using the piece of centimeter graph paper given to you, construct a line graph which will illustrate this data. Answer questions 29 through 34.

29. On which axis did you place the manipulated or independent variable?

    A. X
    B. Y

Appendix A / 233

30. On which axis did you place time?

    A. X
    (B.) Y

31. On which axis did you place the days?

    (A.) X
    B. Y

32. On which axis did you place the responding or dependent variable?

    A. X
    (B.) Y

33. On the basis of the information presented in the graph, how much time do you predict the rat will take to travel the maze on Day 7?

    (A.) 18 seconds
    B. 20 seconds
    C. 22 seconds
    D. 24 seconds

34. What do you estimate the rat's time was on Day 4.5?

    A. 34.5 seconds
    B. 36.0 seconds
    (C.) 38.0 seconds
    D. 40.0 seconds

MODULE 8
INTERPRETING DATA

The mass of an object was measured independently by ten different persons with the same set of gram masses and the same equal-arm balance. Their measurements are shown below.

Person who made observation	Observed mass of objects (*in grams*)
Hal	9.0
Bill	10.0
Sally	10.5
Al	9.5
Jerry	10.0
John	12.0
Shari	11.0
Charlette	10.5
Fran	10.0
Pete	7.5

35. On the basis of the data, what is your best estimate of the mass of the object?

    A.   9.0 g
    B.   9.5 g
    (C.) 10.0 g
    D.   10.5 g

36. What is the median value?

    A.   9.5
    (B.) 10.0
    C.   10.5
    D.   11.0

37. What was the range of the values?

    (A.) 4.5 g
    B.   7.5 g
    C.   10.0 g
    D.   11.0 g

38. Which is *not* a valid inference?

    A.   Best answer for object's mass is 10.0 g $\pm$ 0.5 g.
    B.   John is not very accurate in measuring.
    C.   Pete does careless work.
    (D.) The mode is 10.0 g.

MODULE 9
CONTROLLING VARIABLES

It is reported that Galileo, while attending Mass in the Cathedral of Pisa, noticed a swinging chandelier and became interested in the time it took for the chandelier to move from one end of its swing to the other and back to its original position.

In answering the following questions, consider a pendulum to be an object (the pendulum bob) suspended by a string (or other means of support). The time for a complete swing of a pendulum is called the *period* of the pendulum.

Answer questions 39 through 41 which are based on the two paragraphs that you have just read.

39. Which variable would *not* affect the period of a pendulum?

    A. Distance of swing
    B. Length of the string
    C. Mass of the bob
    D. Type of support

40. Is the period a variable?

    A. No
    B. Yes

41. If the period is a variable, what kind of variable is it?

    A. Manipulated
    B. Responding
    C. Not a variable

You have planted some bean seeds under identical conditions with one exception. You are giving the plants various measured amounts of water. There is a resulting difference in growth among the bean plants. Answer the following two questions which are concerned with this activity.

42. The amount of water given to the plants is an example of what kind of variable?

    (A.) Manipulated
    B. Responding
    C. Both
    D. Neither

43. The resulting difference in growth is an example of what kind of variable?

    A. Manipulated
    (B.) Responding
    C. Both
    D. Neither

MODULE 10
EXPERIMENTING — FORMULATING HYPOTHESES

Suppose your class has been experimenting for several weeks with the growth and development of green plants. One of the students observes (among other things) that plants in the classroom move or bend toward the windows. The student infers that the green plants in the classroom bend toward the windows in response to light. On the basis of this information, answer items 44 through 48.

44. Select the best choice of a hypothesis constructed from the student's inference.

    A. All stems bend toward a light source.
    B. Green stems always bend toward windows.
    (C.) The stems of green plants grow toward a light source.
    D. Ultra-violet light helps plants grow.

45. Select the experiment that would test the hypothesis stated in the last item.

    A. Grow a set of established seedlings in light. Then place the set in a dark place. Periodically test for any curvature of the stems.

B. Grow a set of established seedlings in light. Then place the set in the end of a box that has a single hole in the opposite end. Allow light to pass through. Periodically test for any curvature of the stems.

C. Grow two sets of three kinds of established seedlings in light. Then place one set in a dark place and the other set in the end of a box that has a single hole in the opposite end. Allow light to pass through the hole. Periodically measure any curvature of the stems.

46. What important experimental procedure was omitted from one or more of the above investigations?

A. Control
B. Fertilizer
C. Light
D. Water

47. Which of the following observations resulting from the test would support the hypothesis?

A. The stems in the box bend toward the opening
B. The stems in the dark do not bend from the vertical position
C. The combination of A and B
D. None of the above

48. Which of the observations resulting from the test would *not* support the hypothesis?

A. The plants in the box do not bend toward the light
B. The plants in the box bend away from the light
C. Both groups of plants bend, or neither group bends
D. All of the above

MODULE 11
DEFINING OPERATIONALLY

Do the following three test items.

49. Which of the following is a more appropriate operational definition of *oxygen?*

 (A.) A gas that causes a glowing splint to burst into flame when the splint is placed into a container of the gas
 B. An element composed of atoms having atomic number 8 and atomic weight 16

50. Which of the following is a better operational definition of *density?*

 A. A relation between the mass and volume of an object
 (B.) The quotient of the mass of an object in grams and the volume of the object in milliliters.

51. Choose the best operational definition of *good conductor of electricity.*

 A. A good conductor of electricity is any object which can be used to complete an electrical circuit.
 (B.) A wire is removed from an arrangement of batteries, wires, and light bulb in which the bulb is glowing, and the light bulb goes out; if an object is used to replace the wire and the light goes on again, the object is a good conductor of electricity.
 C. A good conductor of electricity is any object through which electricity flows easily.

MODULE 12
INSTRUCTIONAL OBJECTIVES

Read the following paragraph and answer item 52.

A child is being observed in the classroom. The child has been given a geometrical shape cut from construction paper. The teacher asks: Is this object symmetrical? The child responds: Yes, it is. The teacher inquires:

Appendix A / 239

How do you know it is symmetrical? In response, the child does something with the geometrical shape which indicates how he decides whether the shape is symmetrical.

52. Which action word satisfactorily names the performance of the child in response to the question?

    A. Constructs
    B. Demonstrates
    C. Identifies
    D. Names

For items 53 through 55, read each question and choose one of the answer choices listed below it.

53. Which of the following is the conclusion of a behavioral objective that begins like this: *Upon completion of this exercise the child should be able to:*

    A. Construct a grouping of living and nonliving objects on the basis of observable properties with at least 90% accuracy.
    B. Communicate clearly.
    C. Understand how to infer the contents of a packaged article.

54. Which conclusion will give the behavioral objective the highest level of difficulty? Upon completing this activity, the student should be able to recognize and name:

    A. Changes in such characteristics as temperature, size, shape, and color observed in solid-liquid changes.
    B. Color changes.
    C. The primary and secondary colors.
    D. Two or more characteristics of an object such as color, size, shape, and texture.

55. Which of the following statements about behavioral objectives is true?

    A. Performance objectives are useful in the measurement of understandings, appreciations, and the long-range goals of science.

B. An observer is usually required as a witness in order to determine if the learner has acquired the desired behavior.

C. All behavioral objectives call for student action.

D. The performance objectives that involve applications of concepts are more easily forgotten than simple recall and manipulative objectives.

If you have any time remaining, go back over the test items to be sure you have darkened the space on the answer card that corresponds to your choice for an answer to each test item.

Appendix
Where do we go from here?

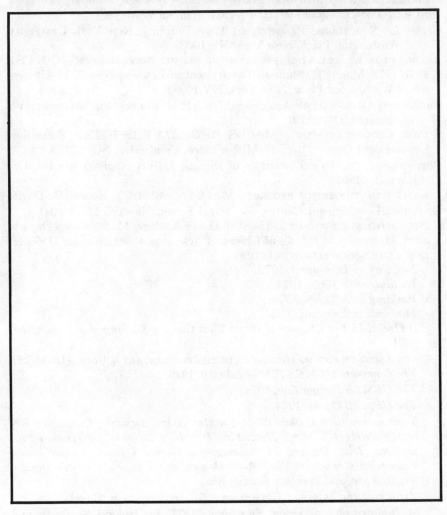

Sources for more ideas and activities

FOR TEACHERS

SOURCES FOR COURSE AND CURRICULUM IMPROVEMENT PROJECTS

Biological Sciences Curriculum Study Company, Box 930, Boulder, CO 80302
HUMAN SCIENCES PROGRAM (HSP) 1971. William V. Mayer.

Center for Educational Research, 51 Press Building, New York University, 32 Washington Place, New York, NY 10003
CONCEPTUALLY ORIENTED PROGRAM IN ELEMENTARY SCIENCE (COPES) 1967–1973. Morris H. Shamos, Department of Physics, New York University, 4 Washington Place, New York, NY 10003

Curriculum Development Associates, Inc., 1211 Connecticut Avenue NW, Washington, DC 20036
MAN: A COURSE OF STUDY (MACOS) 1963–1970. Peter B. Dow, Education Development Center, Inc., 15 Mifflin Place, Cambridge, MA 02138

Curriculum Laboratory, University of Illinois, 1210 Springfield Avenue, Urbana, IL 61801
MADISON MATHEMATICS PROJECT (MADM) 1961–1973. Robert B. Davis.

Education Development Center, 55 Chapel Street, Newton, MA 02160
THE ARITHMETIC PROJECT 1965–1970. David A. Page, Mathematics Department, University of Illinois at Chicago Circle, Box 4348, Chicago, IL 60680
EARLY CHILDHOOD EDUCATION STUDY
Building a Playground, 1970.
Building with Tires, 1971.
Building with Tubes, 1970.
Moments in Learning, 1968.
A Useful List of Classroom Items That Can Be Scrounged or Purchased, 1971.

UNIFIED SCIENCE AND MATHEMATICS IN THE ELEMENTARY SCHOOL (USMES)
The Complete USMES School Library, 1975.
The USMES Design Lab, 1974.
The USMES Guide, 1974.
Teacher resource books, 1974: *Burglar Alarm Designs, Consumer Research Product Testing, Describing People, Designing for Human Proportions, Dice Design, Electromagnetic Device Design, Lunch Lines, Pedestrian Crossings, Play Area Design and Use, Soft Drink Design, Traffic Flow,* and *Weather Predictions.*
Arranging the Informal Classroom, 1973, by Brenda S. Engel.
A Bibliography of Open Education, 1971, by Roland S. Barth and Charles H. Rathbone.
Children's Literature: A Bibliography, 1972.

A Classroom for Young Children: Approximation No. 1, 1966, by Allan Leitman and Edith H. F. Churchill.

Infant School, 1969, by Courtney B. Cazden.

An Interview with Bruce Whitmore, 1969.

An Interview with Dorothy Welch, 1969.

An Interview with Pat Hourihan, 1971.

Environmental Studies for Urban Youth, Evergreen State College, Olympia, WA 98505

ENVIRONMENTAL STUDIES FOR URBAN YOUTH (ES) 1970. Richard R. Sluss.

Ginn and Company, Xerox Education Group Distribution Center, 555 Gotham Parkway, Carlstadt, NJ 07072

SCIENCE — A PROCESS APPROACH (SAPA) Commission on Science Education, 1962–1971.

Harper and Row, Inc., Elementary and High School Division, 2500 Crawford Avenue, Evanston, IL 60201

ELEMENTARY SCHOOL SCIENCE PROJECT (ESSP) 1960–1969. J. Myron Atkin, College of Education, and Stanley P. Wyatt, Department of Astronomy, University of Illinois, Urbana, IL 61801.

D. C. Heath and Company, 285 Columbus Avenue, Boston, MA 02116

UNIVERSITY OF ILLINOIS COMMITTEE ON SCHOOL MATHEMATICS (UICSM) 1962–1971. Russell E. Zwoyer, University of Illinois Curriculum Laboratory, 1210 West Springfield, Urbana, IL 61801

Houghton Mifflin Company, One Beacon Street, Boston, MA 02108

INVESTIGATING THE EARTH: EARTH SCIENCE CURRICULUM PROJECT (ESCP) 1963–1973. William D. Romey, Box 1559, Boulder, CO 80302

Hubbard Scientific Company, PO Box 10, Northbrook, IL 60002

BIOLOGICAL SCIENCES CURRICULUM STUDY (BSCS) 1972–1976. Life sciences for the educable mentally handicapped.

Imperial International Learning, Kankakee, IL 60901

INDIVIDUALIZED SCIENCE 1968–1972. Leo Klopfer et al. University of Pittsburgh, College of Education, Pittsburgh, PA

Institute for Mathematical Studies in the Social Sciences, Stanford University, Stanford, CA 94305

EXPERIMENTAL TEACHING OF MATHEMATICS IN THE ELEMENTARY SCHOOL 1959–1971. Patrick Suppes.

ISIS Headquarters, 415 North Monroe Street, Tallahassee, FL 32301

INDIVIDUALIZED SCIENCE INSTRUCTIONAL SYSTEM (ISIS) 1972–. Ernest Burkman, Florida State University, Tallahassee, FL 32306

J. B. Lippincott Company, East Washington Square, Philadelphia, PA 19105

BSCS ELEMENTARY SCHOOL SCIENCE PROGRAM 1975 (grades 4–6); 1978 (grades K–3).

Lawrence Hall of Science, University of California, Berkeley, CA 94720

OUTDOOR BIOLOGY INSTRUCTIONAL STRATEGIES (OBIS) 1972–. Watson M. Laetsch.

OBIS Trial Editions; Outdoor Biology Instructional Strategies.

SCIENCE CURRICULUM IMPROVEMENT STUDY (SCIS)

SCIS Evaluation Supplements for *Communities, Ecosystems, Electric and Magnetic Interactions, Energy Sources, Environments, Interaction and Systems, Life Cycles, Material Objects, Populations, Relative Position and Motion, Organisms,* and *Subsystems and Variables.*

SCIS Omnibus (a collection of Readings from 1962–1973 related to SCIS), 1973.

SCIS Teacher's Handbook, 1974.

Teacher's Guide for Science for Kindergarten, 1974.

McGraw-Hill Book Company, Webster Division, 1221 Avenue of the Americas, New York, NY 10020

ELEMENTARY SCIENCE STUDY (ESS) 1962–1973. Joseph Griffith, Education Development Center, 55 Chapel Street, Newton, MA 02160

The ESS Reader, 1970.

A Materials Book for the Elementary Science Study, 1972.

TIME, SPACE, AND MATTER: SECONDARY SCHOOL SCIENCE PROJECT (TSM) 1963–1972. George J. Pallrand, Science Education Center, Rutgers, The State University, New Brunswick, NJ 08903.

McGraw-Hill-Ryerson Ltd., Toronto, Canada

ELEMENTARY SCIENCE CURRICULUM STUDY (ESCS). Robert K. Crocker.

ESCS Teaching Guide, Volume 1 (grades 1–3), 1973.

ESCS Teaching Guide, Volume 2 (grades 4–6), 1973.

Bicycles to Beaches, 1972, by W. Gillespie et al.

Broadway to Boot Hill, 1974, by W. Gillespie et al.

National Association of Geology Teachers, Department of Earth Sciences, Southeast Missouri State University, Cape Girardeau, MO 63701

CRUSTAL EVOLUTION EDUCATION PROJECT 1976–80. Edward C. Stoever, Jr.

Prentice-Hall, Inc., Englewood Cliffs, NJ 07632

INTRODUCTORY PHYSICAL SCIENCE (IPS) 1963–1969. Uri Haber-Schaim, Education Development Center, Inc., 55 Chapel Street, Newton, MA 02160

Project on Elementary School Mathematics and Science, c/o Peter B. Shoreman, College of Education, University of Illinois, Urbana, IL 61801

PROJECT ON ELEMENTARY SCHOOL MATHEMATICS AND SCIENCE (PESMS) 1969–1973.

Rand-McNally and Company, Box 7600, Chicago, IL 60680

SCIENCE CURRICULUM IMPROVEMENT STUDY (SCIS)

Teacher's guides for 12 booklets; pupil activity books.

W. B. Saunders Company, West Washington Square, Philadelphia, PA 19105

MINNESOTA SCHOOL OF MATHEMATICS AND SCIENCE PROJECT (MINNEMAST) 1961–1970. James H. Werntz, Jr., University of Minnesota, Minnesota School of Mathematics and Science Center, 720 Washington Avenue SE, Minneapolis, MN 55455

School Mathematics Study Group, c/o E. G. Begle, School of Education, Stanford University, Stanford, CA 94305

SCHOOL MATHEMATICS STUDY GROUP (SMSG) 1958–1972.

Silver Burdett Company, Morristown, NJ 07960

NATIONAL ENVIRONMENTAL EDUCATION DEVELOPMENT PROGRAM (NEED), Educational Challenges, Inc., Silver Burdett General Learning Corporation.

PROBING THE NATURAL WORLD: INTERMEDIATE SCIENCE CURRICULUM STUDY (ISCS) 1969–. David Redfield and William R. Snyder, Florida State University, Tallahassee, FL 32306

Level III minicourses for the middle grades (6 and up) including a teacher's edition, a student record book, and a master set of equipment: *Crustry Problems, Environmental Science, Investigating Variation, In Orbit, Well-Being, What's Up? Why You're You*, and *Winds and Weather*.

SOURCES FOR LEARNING CENTER IDEAS

Addison-Wesley Publishing Company, Inc., Sand Hill Road, Menlo Park, CA 95025
Developmental math cards.

Allyn and Bacon, Inc., 470 Atlantic Avenue, Boston, MA 02210
Investigate and Discover: Elementary Science Lessons, 1975, by Robert B. Sund, Bill W. Tillery, and Leslie W. Trowbridge.
Involving Students in Questioning, 1976, by Francis P. Hunkins.

Tunis Baker, 650 Concord Drive, Holland, MI 49423
Baker Science Packets.

Contemporary Ideas, Box 1703, Los Gatos, CA 95030
Math activities, books, and assorted math game packets.

Creative Publications, Box 328, Palo Alto, CA 94320
Creative math activities.

CTM Publishing Company, Box 1513, Sunnyvale, CA 95088
Learning Center Guide, 1972, by Barbara Guiske and Bernard T. Cote.

Dell Publishing Company, 750 Third Avenue, New York, NY 10017
Big Rock Candy Mountain, 1971, edited by Samuel Yanes and Cia Holdorf.

Educational Resources Information Center (ERIC) for Science, Mathematics, and Environmental Education, 1800 Cannon Drive, 400 Lincoln Tower, Ohio State University, Columbus, OH 43210
100 Teaching Activities on Environmental Education, 1974, by John H. Wheatley and Herbert L. Coon.

ERIC Clearinghouse on Early Childhood Education, 805 West Pennsylvania Avenue, Urbana, IL 61801
Opening up the Classroom: A Walk Around the School, 1971, by Sylvia Hucklesby.

Educational Service, Inc., Box 219, Stevensville, MI 49127
Nine elementary teacher's aid books with excellent ideas.

Fearon Publishers, 6 Davis Drive, Palo Alto, CA 94002
Preparing Instructional Objectives, 1962, by Robert F. Mager.

Follett Publishing Company, 1010 West Washington Boulevard, Chicago, IL 60607

Individualizing Instruction and Keeping Your Sanity, 1973, by William M. Bechtol.

Junior activity books.

Garrard Publishing Company, 1607 North Market Street, Champaign, IL 61920

Reading games and language arts activities.

Harcourt Brace Jovanovich, Inc., 757 Third Avenue, New York, NY 10017

Teaching Elementary Science Through Investigation and Colloquium, 1971, by Brenda Landsdown, Paul E. Blackwood, and Paul F. Brandwein.

Harper and Row, Inc., 10 East 53rd Street, New York, NY 10022

Classroom Questions: What Kinds? 1966, by Norris M. Sanders.

Highlights for Children, 2300 West Fifth Avenue, Columbus, OH 43216

Fun/do packs containing a variety of activities.

Holt, Rinehart, and Winston, Inc., 383 Madison Avenue, New York, NY 10017

Environmental Education in the Elementary School, 1972, by Larry L. Sale and Ernest W. Lee.

Teachers, Children, and Things: Materials-Centered Science, 1971, by Clifford J. Anastasion.

Ideal School Supply Company, 11002 South Lavergne Avenue, Oak Lawn, IL 60453

Laminated crossword puzzles, magic cards, and number puzzles.

Incentive Publishing Company, Box 12522, Nashville, TN 37212

Center Stuff for Nooks, Crannies, and Corners: Complete Learning Centers for the Elementary Classroom, 1973, by Imogene Forte et al.

Cornering Creative Writing: Learning Centers, Games, Activities, and Ideas for the Elementary Classroom, 1973, by Imogene Forte et al.

Creative Math Experiences for the Young Child, 1973, by Imogene Forte and Joy MacKenzie.

Creative Science Experiences for the Young Child, 1973, by Imogene Forte and Joy MacKenzie.

Nooks, Crannies, and Corners: Learning Centers for Creative Classrooms, 1972, by Imogene Forte et al.

Individualized Books Publishing Company, Box 591, Menlo Park, CA 94025

Individualize with Learning Station Themes, 1973, by Lorraine Godfrey.

Individualizing Through Learning Stations, 1972, by Lorraine Godfrey.

Instructor Publications, Inc., Dansville, NY 14437

Metric Measurement Activities and Bulletin Boards, 1973, by Cecil R. Trueblood.

Marie's Educational Materials, 195 South Murphy Avenue, Sunnyvale, CA 94086

Variety of reading and math aids.

Charles E. Merrill Publishing Company, 1300 Alum Creek Drive, Columbus, OH 43216

Behavioral Objectives and Evaluational Measures, Science and Mathematics, 1972, by Robert B. Sund and Anthony J. Picard.

Mind/Matter Corporation, Box 345, Danbury, CT 06810.
  Excellent manipulative and motivational materials.
National Aeronautics and Space Administration (NASA), Educational Publications, Washington, DC 20546
National Science Teachers Association, 1742 Connecticut Avenue NW, Washington, DC 20009
  "Outstanding Science Trade Books for Children," *Science and Children.*
  Metric Exercises: Lively Activities on Length, Weight, Volume, and Temperature, 1973. Stock #471–14664.
  Elementary Science Packets on the following topics: Drug Education; Environmental Education — I. Environmental Education — II. Measurement and the Metric System; Multidisciplines.
Parker Publishing Company, West Nyack, NY 10994
  Individualized Techniques for Teaching Earth Sciences, 1975, by Joseph D. Exline.
  Outdoor Science for the Elementary Grades, 1972, by John H. Rosengren.
Pawnee Publishing Company, 1 Pondfield Road, Bronxville, NY 10708
  Investigating Air, Land, and Water Pollution, 1971, by Diane Storin.
Prentice-Hall, Inc., Englewood Cliffs, NJ 07632
  Classroom Instructional Tactics, 1973, by W. James Popham and Eva L. Baker.
  Developing Teacher Competencies, 1971, edited by James E. Weigand.
  Evaluating Instruction, 1973, by W. James Popham.
Teachers College Press, Columbia University, 1234 Amsterdam Avenue, New York, NY 10027
  The Experience of Science: A New Perspective for Laboratory Teaching, 1976, by O. Roger Anderson.
Teachers Exchange of San Francisco, 600 35th Avenue, San Francisco, CA 94121
  Excellent math task cards and a variety of other materials.
U.S. Department of the Interior, Bureau of Land Management, Washington, DC 20240
  All Around You: An Environmental Study Guide, 1971, Superintendent of Documents, Stock #2411–0043.

PERIODICALS

Learning. 1255 Portland Place, Boulder, CO 80302
School Science and Mathematics. School Science and Mathematics Association, Lewis House, PO Box 1614, Indiana University of Pennsylvania, Indiana, PA 15701
Science Activities. 400 Albemarle Street NW, Washington, DC
Science and Children. National Science Teachers Association, 1742 Connecticut Avenue NW, Washington, DC 20009

Science Education. John Wiley and Sons, 605 Third Avenue, New York, NY 10016

Science Teacher. National Science Teachers Association, 1742 Connecticut Avenue, NW, Washington, DC 20009

BOOKS

A Child's Garden: A Guide for Parents and Teachers, 1972. Chevron Chemical Company, Public Relations, 200 Bush Street, San Francisco, CA 94120

Children and Science, 1975, by David P. Butts, and Gene E. Hall. Prentice-Hall, Inc., Englewood Cliffs, NJ 07632

City Planning: The Games of Human Settlement, 1975, by Forrest Wilson. Van Nostrand Reinhold Company, 450 West 33rd Street, New York, NY 10001

Creative Activities Resource Book for Elementary School Teachers, 1978, by Thomas Turner. Reston Publishing Company, Reston, VA 22090

Early Childhood Curriculum: A Piaget Program, 1970, by Celia Stendler Lavatelli. American Science and Engineering, Inc., 20 Overland Street, Boston, MA 02215

Energy Environment Source Book, 1975, by John M. Fowler. National Science Teachers Association, 1742 Connecticut Avenue NW, Washington, DC 20009

Energy: Historical Development of the Concepts, 1975, by R. Bruce Lindsay. Dowden, Hutchinson and Ross, Box 699, Stroudsburg, PA 18360

Energy: Resource, Slave, Pollutant, 1975, by Robert Rouse and Robert O. Smith. Macmillan, Inc., 866 Third Avenue, New York, NY 10022

Energy: The Continuing Crisis, 1977, by Norman Metzger. Crowell Press, New York, NY

Energy: The Solar Prospect, 1977, by Denis Hayes. Worldwatch Institute, 1776 Massachusetts Avenue NW, Washington, DC

Errors in Experimentation, 1977, by Carl W. Hall. Matrix Publishers, Champaign, IL 61820

Evolutionary Ecology, 1978, by Erik R. Pianka. Harper and Row, Inc., 10 East 53rd Street, New York, NY 10022

Games for the Science Classroom, 1977, by Paul B. Hounshell and Ira Trollinger. National Science Teachers Association, 1742 Connecticut Avenue NW, Washington, DC 20009

How Children Learn Science: Conceptual Development and Implications for Teaching, 1977, by Ronald G. Good. Macmillan, Inc., 866 Third Avenue, New York, NY 10022

Living: An Attitude of Imagination, 1976, by Pam Early. Kendall/Hunt Publishing Company, Dubuque, Iowa 52001

Living with Energy, 1978, by Robert Alves. Viking Press, 625 Madison Avenue, New York, NY 10022

Magic, Science and Civilization, 1978, by Jacob Bronowski. Columbia University Press, 562 West 113th Street, New York, NY 10025

Making Scientific Toys, 1975, by Carson Ritchie. Nelson Publishing Company, Nashville, TN 37203

On Aesthetics in Science, 1978, by Judith Wechsler. MIT Press, 28 Carleton Street, Cambridge, MA 02142

New UNESCO Source Book for Science Teaching, 1973. UNIPUB, Inc., 650 First Avenue, New York, NY 10017

Principles of Three-Dimensional Design, 1977, by Wucius Wong. Van Nostrand Reinhold Company, 450 West 33rd Street, New York, NY 10001

Progress and Its Problems: Toward a Theory of Scientific Growth, 1977, by Larry Laudan. University of California Press, 2223 Fulton Street, Berkeley, CA 94720

Quasar, Quasar, Burning Bright, 1978, by Isaac Asimov. Doubleday Publishing Company, 245 Park Avenue, New York, NY 10017

Readings in Science Education for the Elementary School, 1975, by Edward Victor and Marjorie S. Lerner. Macmillan, Inc., 866 Third Avenue, New York, NY 10022

Science and Building, 1978, by Henry J. Cowan. Wiley Interscience Publications, 605 Third Avenue, New York, NY 10016

Science and Creation in the Middle Ages, 1976, by Nicholas Steneck. University of Notre Dame Press, Notre Dame, IN

Science and Immortality, 1977, by William Osler. Arno Press, 330 Madison Avenue, New York, NY 10017

Science and Its Critics, 1978, by John Arthur Passmore. Rutgers University Press, New Brunswick, NJ 08901

Science and Its Public: The Changing Relationship, 1976, edited by Gerald Holton and William A. Blanpied. D. Reidel Publishing Company, 306 Dartmouth Street, Boston, MA 02116

Science and Society: Past, Present and Future, 1975, by Nicholas Steneck. University of Michigan Press, Ann Arbor, MI 48106

Science as a Human Endeavor, 1978, by George F. Kneller. Columbia University Press, 562 West 113th Street, New York, NY 10025

Science Development: Toward the Building of Science in Less Developed Countries, 1975, by Michael J. Morovesik. Indiana University Press, Bloomington, IN 47401

Science Experiences for Young Children, 1975, by Rosemary Althouse and Cecil Main. Teachers College Press, 1234 Amsterdam Avenue, New York, NY 10027

Science Experiences for Young Children, 1975, by Viola Carmichael. Southern California Association for the Education of Young Children Press, Los Angeles, CA

Science Fair Project Index, 1960–72, 1975, edited by Janet Y. Stoffer. Scarecrow Press, 52 Liberty Street, Metuchen, NJ 08840

Science for the Elementary School, 1975, by Edward Victor. Macmillan Inc., 866 Third Avenue, New York, NY 10022

Science Since Babylon, 1975, by Derek De Solla Price. Yale University Press, New Haven, CT 06511

Science, Technology and the Environment, 1975, by John T. Hardy. W. B. Saunders Company, West Washington Square, Philadelphia, PA 19105

Science: Who Needs It? 1975, by Ben Bova. Westminster Press, Witherspoon Building, Philadelphia, PA 19107

Sourcebook for Biological Sciences, 1972, by Donald L. Troyer, Maurice G. Kellogg, and Hans O. Andersen. Macmillan, Inc., 866 Third Avenue, New York, NY 10022

Sourcebook for Earth Sciences and Astronomy, 1972, by Russell O. Utgard, George T. Ladd, and Hans O. Andersen. Macmillan, Inc., 866 Third Avenue, New York, NY 10022

The Ecology of Man: An Ecosystem Approach, 1976, by Robert Leo Smith. Harper and Row, Inc., 10 East 53rd Street, New York, NY 10022

The Game of Science, 3rd ed., 1977, by Garvin McCain and Erwin M. Segal. Wadsworth Publishing Company, Belmont, CA 94002

The Great Perpetual Learning Machine, 1976, by Jim Blake and Barbara Ernst. Little, Brown and Company, 34 Beacon Street, Boston, MA 02106

The Science of Ethics of Equality, 1977, by David Hawkins. Basic Books, 10 East 53rd Street, New York, NY 10022

The Spheres of Life: An Introduction to World Ecology, 1975 by Joseph W. Meeker. Charles Scribner's Sons, 597 Fifth Avenue, New York, NY 10017

Toward One Science: The Convergence of Traditions, 1978, by Paul Snyder. St. Martin's Press, 175 Fifth Avenue, New York, NY 10010

Water Mix Experiments Teacher's Guide: Conceptually Oriented Program in Elementary Science (COPES), 1971, American Science and Engineering, Inc., 20 Overland Street, Boston, MA 02215

Weather Study: An Approach to Scientific Inquiry, 1972, by J. W. Bainbridge and R. W. Stockdale. Methuen Educational Ltd., 11 New Fetter Lane, London EC 4, England.

Will the Real Teacher Please Stand Up? 1972, by Mary Greer and Bonnie Rubenstein. Goodyear Publishing Company, 15115 Sunset Boulevard, Pacific Palisades, CA 90272

FOR TEACHERS AND CHILDREN

BOOKS

Addison-Wesley Publishing Company, Inc., Reading, MA 01867
  Investigating Science with Coins, 1969, by Laurence B. White, Jr.
  Investigating Science with Nails, 1969, by Laurence B. White, Jr.
  Investigating Science with Paper, 1969, by Laurence B. White, Jr.
  Investigating Science with Rubber Bands, 1969, by Laurence B. White, Jr.
  Science Games, 1975, by Laurence B. White, Jr.

Science Series for the Young, 1969, by Herbert H. Wong and Matthew F. Vessel. Including *Our Tree, My Ladybug, My Goldfish,* and *Our Terrariums.*

Science Tricks, 1975, by Laurence B. White, Jr.

The Riddle of the Stegosaurus, 1969, by D. C. Ipsen.

What Does a Bee See?, 1971, by D. C. Ipsen.

Animal Care From Protozoa to Small Mammals, 1977, by F. Barbara Orlans.

Science Puzzles, 1975, by Laurence B. White.

Science Toys, 1975, by Laurence B. White.

American Education Publications, 245 Long Hill Road, Middletown, CT 06457

Clowns, Colors, Corks, and Carrots, 1972, by John F. Mongillo.

Art Education, Inc., Blauvelt, NY 10913

100 Ways to Have Fun with an Alligator and 100 Other Involving Art Projects, 1969, by Norman Laliberte and Richey Kehl.

Atheneum Publishers, 122 East 42nd Street, New York, NY 10017

What Is It Really Like Out There? Objective Knowing, 1977, by Thomas Moorman.

Belwin-Mills Publishing Corporation, 250 Maple Avenue, Rockville Centre, NY 11570

The Sierra Club Survival Songbook, 1971, by Jim Morse and Nancy Mathews.

Marshall Cavendish Corporation, 110 East 54th Street, New York, NY 10022

The Illustrated Encyclopedia of Science and Technology: How It Works, 1977.

Creative Publications, Inc., Box 10328, Palo Alto, CA 94303

Georule Activities, 1971, by Ernest R. Ranucci.

Pattern Blocks Coloring Books, 1974, by Linda Silvey and Marion Pasternack.

Pic-a-Puzzle, 1970, by Reuben A. Schadler and Dale G. Seymour.

String Sculpture, 1972, by John Winter.

Thomas Y. Crowell Company, 10 East 53rd Street, New York, NY 10022

Curiosities of the Cube, 1977, by Ernest R. Ranucci and Wilma E. Rollins.

John Day Company, 666 Fifth Avenue, New York, NY 10019

Food, 1977, by Irving Adler.

Dodd, Mead and Company, 79 Madison Avenue, New York, NY 10016

How to Build a Better Mousetrap Car — And Other Experimental Science Fun, 1977, by Al G. Renner.

Doubleday and Company, 245 Park Avenue, New York, NY 10017

More Brain Boosters, 1975, by David Webster.

More Brain Boosters, with photos, 1975, by David Webster.

Dover Publications, 180 Varick Street, New York, NY 10014

Soap-Bubbles, 1959, by C. V. Boys.

M. Evans and Company, 216 East 49th Street, New York, NY 10017

The Good Drug and the Bad Drug, 1970, by John S. Mars, M.D.

Farallones Designs, Star Route, Point Reyes Station, CA 94956
  Making Places, Changing Spaces in Schools, at Home, and Within Ourselves, 1971.
Garrard Publishing Company, 1607 North Market Street, Champaign, IL 61820
  Exploring Fields and Lots: Easy Science Projects, 1978, by Seymour Simon
Grosset and Dunlap, 51 Madison Avenue, New York, NY 10010
  "Wonder Starters." Including *Bees, Clothes, Dinosaurs, Eggs, Fire, Hair, Homes, Milk, Rain, Sleep, Teeth,* and *Telephone.*
Houghton Mifflin Company, One Beacon Street, Boston, MA 02108
  Recyclopedia: Games, Science Equipment, and Crafts from Recycled Materials, 1976, Robin Simons.
  Kids are Natural Cooks: Child-Tested Recipes for Home and School Using Natural Foods, 1972, by Parents' Nursery School.
Little, Brown and Company, 34 Beacon Street, Boston, MA 02106
  I Saw a Purple Cow and 100 Other Recipes for Learning, 1972, by Ann Cole.
McDonald's Ecology Action Pack, 1973, PO Box 2344, Kettering, OH 45429
McGraw-Hill Book Company, 330 West 42nd Street, New York, NY 10036
  Colourweeples, 1972, by D. Craig Gillespie.
  Weeple People, 1972, by D. Craig Gillespie.
  McGraw-Hill Dictionary of the Life Sciences, 1976, edited by Daniel Lapedes.
Mira Math Company, PO Box 625, Station B, Willowdale, Ontario, MZK 2PQ, Canada
  Mira Math Activities for the Elementary School, 1973.
  Mira Math Activities for High School, 1973.
C. V. Mosby Company, St. Louis, MO 63141
  Earth in Crisis: An Introduction to the Earth Sciences, 1976, by Thomas L. Burns and Herbert J. Spiegel.
Jeffrey Norton Publishers, 145 East 49th Street, New York, NY 10017
  Deal Me In! The Use of Playing Cards in Teaching and Learning, 1973, by Margie Golick.
Parker Publishing Company, West Nyack, NY 10994
  Illustrated Treasury of General Science Activities, 1975, by Robert G. Hoehn.
Prentice-Hall, Inc., Englewood Cliffs, NJ 07632
  The Centering Book — Awareness Activities for Children, Parents, and Teachers, 1975, by Gay Hendricks.
  Experiments with Everyday Objects, 1978, by Kevin Goldstein-Jackson, Norman Rudnick, and Ronald Hyman.
  Paper, Pencils and Pennies: Games for Learning and Having Fun, 1977, by Ronald T. Hyman.
Random House, 201 East 50th Street, New York, NY 10022
  Charlie Brown's Second Super Book of Questions and Answers about the Earth and Space, 1977.

Scholastic Book Services, Four Winds Press, 50 West 44th Street, New York, NY 10036

Just a Box, 1973, by Goldie Taub Chernoff.

Music and Instruments for Children to Make, 1972, by John Hawkinson and Martha Faulhaker.

Pebbles and Pods, 1972, by Goldie Taub Chernoff.

Play with Paper, 1973, by Thea Bank-Jensen.

Sheed and Ward, Inc., 6700 Squibb Road, Mission, KS 66202

Sneaky Feats: The Art of Showing Off and 53 Ways to Do It, 1975, Tom Ferrell and Lee Eisenberg.

Simon and Schuster, Rockefeller Center, 630 Fifth Avenue, New York, NY 10020

The Great International Paper Airplane Book, 1967, by Jerry Mander et al.

Performing Plants, 1969, by Ware T. Budlong.

Troubador Press, 126 Folsom Street, San Francisco, CA 94105

Geometric Playthings to Color, Cut and Fold, 1973, by Jean J. Pedusen and Kent A. Pedusen.

University of California Press, 2223 Fulton Street, Berkeley, CA 94720

Great Scientists Speak Again, 1975, by Richard M. Eakin.

Van Nostrand Reinhold Company, 450 West 33rd Street, New York, NY 10001

The Collector's Guide to Rocks and Minerals, 1975 by James R. Tindall and Roger Thornhill.

Van Nostrand's Scientific Encyclopedia, 1976, edited by Douglas Considine.

Wayne County Intermediate School District, 1610 Kales Building, Detroit, MI 48226

Suggestions for Getting in Touch with Me and You and Us: For Elementary Teachers and Children, Drug Abuse Reduction Through Education, 1972.

Winston Press, Inc., 25 Groveland Terrace, Minneapolis, MN 55043

Examining Your Environment series, 1972–1974, by J. Kenneth Couchman, John C. MacBean, Adam Stecher, and Daniel F. Wentworth. Includes *Astronomy, Birds, The Dandelion, Ecology, Mapping Small Places, Miniclimates, Pollution, Running Water, Small Creatures, Snow and Ice, Trees*, and *Your Senses*.

FOR CHILDREN

BOOKS

Abelard-Schuman, 666 Fifth Avenue, New York, NY 10019

The Continental Shelves, 1975, by John F. Waters.

Athenum Publishers, 122 East 42nd Street, New York, NY 10017

More About What Plants Do, 1975, by Joan Elma Rahn.
Bantam Books, Inc., 666 Fifth Avenue, NY 10019
   The Lives of a Cell: Notes of a Biology Watcher, 1975, by Lewis Thomas.
Bobbs-Merrill Company, 4300 West 62nd Street, Indianapolis, IN 46206
   The Creation of the Universe, 1977, by David E. Fisher.
Children's Press, 1224 West Van Buren Street, Chicago, IL 60607
   What Will It Be? 1976, Jane Belk Moncure.
   Wind Is Air: A Concept Book, 1975, by Mary Brewer.
Coward, McCann and Geoghegan, 200 Madison Avenue, New York, NY 10016
   Fitting In: Animals in Their Habitats, 1976, by Gilda and Melvin Berger.
Thomas Y. Crowell Company, 666 Fifth Avenue, New York, NY 10019
   Think Metric, 1973, by Franklyn M. Branley.
   Anno's Counting Book (The World of Numbers), 1977, by Mitsumasa
      Anno. Ages 4–10.
   Black Holes, White Dwarfs and Superstars, 1976, by Franklin M. Branley.
   Cancer, rev. ed., 1977, by Alvin and Virginia Silverstein. Ages 8 and up.
   Caves, 1977, by Roma Gans. Ages 4–8.
   Corals, 1976, by Lili Ronai.
   Energy from the Sun, 1976, by Melvin Berger. Ages 4–8.
   How Life Began, 1977, by Irving Adler. Ages 12 and up.
   How Little and How Much: A Book About Scales, 1976, by Franklyn M.
      Branley.
   Hunger on Planet Earth, 1977, by Jules Archer. Ages 12 and up.
   Little Dinosaurs and Early Birds, 1977, by John Kaufmann. Ages 4–8.
   Living Together in Tomorrow's World, 1976, by Jane Werner Watson.
   The March of the Lemmings, 1976, by James Newton, Ages 4–8.
   Medical Center Lab (Scientists at Work Series), 1976, by Melvin Berger.
      Ages 8 and up.
   Shells Are Skeletons, 1977, by Joan Berg Victor. Ages 4–8.
   The Sunlit Sea, 1976, by Augusta Goldin.
   A Walk in the Forest — The Woodlands of North America, 1977, by Albert
      List, Jr., and Ilka List. Ages 10 and up.
   Wild and Woolly Mammoths, 1977, by Aliki. Ages 4–8.
Crown Publishers, One Park Avenue, New York, NY 10016
   A January Fog Will Freeze a Hog and Other Weather Folklore, 1977,
      edited by Hubert Davis.
John Day Company, 666 Fifth Avenue, New York, NY 10019
   Environment, 1976, by Irving Adler.
   Petroleum: Gas, Oil and Asphalt, 1975, by Irving Adler.
Dodd, Mead and Company, 79 Madison Avenue, New York, NY 10016
   Harness the Wind: The Story of Windmills, 1977, by Joseph E. Brown and
      Anne Ensign Brown.
   What Does A Geologist Do? 1977, by R. V. Fodor.
   The World of the Woodlot, 1975, by Thomas Fegely.
Drake Publishers, 801 Second Avenue, New York, NY 10017
   The Weekend Fossil Hunter, 1977, by Jerry C. LaPlante.
Four Winds Press, 50 West 44th Street, New York, NY 10036

Death Is Natural, 1977, by Laurence Pringle.
Harper and Row, Inc., 10 East 53rd Street, New York, NY 10022
  Elephant Seal Island, 1978, Evelyn Shaw.
  A Nest of Wood Ducks, 1978, by Evelyn Shaw.
Holiday House Inc., 18 East 53rd Street, New York, NY 10022
  The Cloud Book, 1975, by Tomie De Paola.
  Living Together in Nature: How Symbiosis Works, 1977, by Jane E. Hartman.
  The Quicksand Book, 1977, by Tomie De Paola.
J. B. Lippincott Company, East Washington Square, Philadelphia, PA 19105
  Space Monsters: From Movies, TV and Books, 1977, by Seymour Simon.
Lothrop, Lee and Shepard Company, 105 Madison Avenue, New York, NY 10016
  Alpha Centaur: The Nearest Star, 1976, by Isaac Asimov.
Macmillan, Inc., 866 Third Avenue, New York, NY 10022
  The Air of Mars and Other Stories of Time and Place, 1976, edited by Mirra Ginsburg.
  Evening Gray, Morning Red: A Handbook of American Weather Wisdom, 1976, by Barbara Wolff.
  The Hidden World: Life under a Rock, 1977, by Laurence Pringle.
McGraw-Hill Book Company, 1221 Avenue of the Americas, New York, NY 10020
  Science Fun for You in a Minute or Two, 1975, by Herman Schneider.
  Science Fun with a Flashlight, 1975, by Herman and Nina Schneider.
William Morrow and Company, 105 Madison Avenue, New York, NY 10016
  Valleys, 1976, by Della Geotz.
  What Makes a Lemon Sour? 1977, by Gail Kay Haines.
Parents' Magazine Press, 52 Vanderbilt Avenue, New York, NY 10017
  How Heredity Works, 1975, by Jeanne Bendick.
Penguin Books, Harmondsworth, England
  The Penguin Book of the Physical World, 1976, edited by Sonia Larkin and Louis Bernbaum.
G. P. Putnam's Sons, 200 Madison Avenue, New York, NY 10016
  Colonizing Space, 1978, by Erik Bergaust.
  I Was Born in a Tree and Raised by Bees, 1977, by Jim Arnosky.
Random House, 201 East 50th Street, New York, NY 10022
  The Berenstain Bears' Science Fair, 1977, by Stanley Berenstain.
Sterling Publishing Company, 419 Park Avenue South, New York, NY 10016
  Discover the Trees, 1977, by Jerry Cowle.
Van Nostrand Reinhold Books, 120 Alexander Street, Princeton, NJ 08450
  The Art of Light and Color, 1972, by Tom Douglas Jones.
  Building with Balsa Wood, 1965, by John Lidstone.
  Building with Cardboard, 1968, by John Lidstone.
  Building with Wire, 1972, by John Lidstone.
  How to Be a Scientist at Home, 1971, by John Tuey and David Wickers.
  How to Find Out About Zoo Animals, 1972, by Barrington Barber and John Eason.

How to Make and Fly Kites, 1972, by Eve Barwell and Conrad Bailey.

How to Make Things Grow, 1972, by David Wickers and John Tuey.

How to Make Your Own Kinetics, 1972, by David Wickers and Sharon Finmark.

New Math Puzzle Book, 1969, by L. H. Longley-Cook.

Photography Without a Camera, 1972, by Patra Holter.

Small Motors You Can Make, 1963, by John Michel.

Teaching Film Animation, 1971, by Yvonne Anderson.

Walker Press, 720 Fifth Avenue, New York, NY 10019

The California Iceberg, 1975, by Harry Harrison.

Franklin Watts, Inc., 730 Fifth Avenue, New York, NY 10019

Coal, 1976, by Betsy Harvey Kraft.

Western Publishing Company, 1220 Mound Avenue, Racine, WI 53404

I Like to See (A Book About the Five Senses), 1973, by Jane Tymms.

INDEX

Page numbers in italic denote materials teachers may duplicate for use in their classrooms.

Discipline problems, 164. *See also* Behavior problems
Discussions, changing, 39–40
Drugs, 186–187

Educational Policies Commission, 22
Education for All Handicapped Children Act (PL 94-142), 180, 184
Elementary school, science in, 2–3, 132. *See also* Primary grades
Elementary Science Study (ESS), 20–21
Equipment, 24, 163–165. *See also* Materials
  in creative sciencing laboratory, 50–51
  essential, 163
  substitutions for, 52
Ethical issues, 185–191
Evaluation, 206–219
  affective, 213, 218
  cognitive, 206–208
  of creative-sciencer, 220–221
  psychomotor, 219
  by students, 38
  teacher, 208–209, *210–212*
Evolution, 188–189
Exceptional children, 179–184. *See also* Special-education children
  resources on, 184
Experimentation, 156–158

Field trips, 168–171
Film loops, 61
  in laboratory investigations, 49–50
  student-prepared, 50, 61
Films, sound, 62–63
First day of school, tips for, 152–154
"Fourth-grade slump," 40

Gagné, R. M., 18
Games, science, 105–109, 125

Geography, 143
Gifted children
  characteristics of, 177
  defined, 176–177
  science instruction and, 178

Happy Notes, 213, *217*
Health, science and, 144–145

Individualized Educational Program (IEP), 180–181
  sample, 182–183
Individualized instruction, 54–59
  criticism of, 59
  student's role in, 57–58
Inquiry learning, 35
Instructional materials center, 57
Integrated curriculum, 132–150
  vs. isolated, 4
  science in the 4–6, 132–150
Intellectually gifted children, science and, 176–178
Interest centers, 122–126
  sample, 125–126
  types of, 122–123
International Clearinghouse on Science and Mathematics Curricular Developments, 17–18
Inventiveness, 203, 213
Invitations to investigate, 86–89, 126
Involvement, 12
Iowa Tests of Basic Skills, 56

Jargon, educational, 29

Keysort Kid Cards, 59–61

Laboratory, 47–53
  investigations, 48–49
  student guides in, 57
Language arts, science and, 146–150
Learning, 35–36
  individualized, 54–59
Learning categories, 9–10
Learning contracts, student, 56, *58*, 59

Index / 260